SUCCESS THROUGH FAILURE

OTHER BOOKS BY HENRY PETROSKI

Pushing the Limits: New Adventures in Engineering

Small Things Considered: Why There Is No Perfect Design

Paperboy: Confessions of a Future Engineer

The Book on the Bookshelf

Remaking the World: Adventures in Engineering

Invention by Design: How Engineers Get from Thought to Thing

Engineers of Dreams: Great Bridge Builders
and the Spanning of America

Design Paradigms: Case Histories of Error and Judgment
in Engineering

The Evolution of Useful Things

The Pencil: A History of Design and Circumstance

Beyond Engineering: Essays and Other Attempts to Figure
without Equations

To Engineer Is Human: The Role of Failure in Successful Design

SUCCESS THROUGH FAILURE

The Paradox of Design

Henry Petroski

PRINCETON UNIVERSITY PRESS ◈ *Princeton and Oxford*

British Library Cataloging-in-Publication Data is available

Library of Congress Cataloging-in-Publication Data

Petroski, Henry.
Success through failure : the paradox of design / Henry Petroski.
p. cm.
Includes bibliographical references and index.
ISBN-13: 978-0-691-12225-0 ((hardcover) : alk. paper)
ISBN-10: 0-691-12225-3 ((hardcover) : alk. paper)
1. Engineering design—Case studies. 2. System failures
(Engineering)—Case studies. I. Title.
TA174.P4739 2006
620'.0042—dc22   2005034126

Small portions of this material appeared first in *American Scientist,
Harvard Business Review*, and the *Washington Post Book World*

This book has been composed in Adobe Caslon and Helvetica Neue

Printed on acid-free paper. ∞

pup.princeton.edu

Printed in the United States of America

1 3 5 7 9 10 8 6 4 2

To Karen

CONTENTS

PREFACE

This book was written in conjunction with my preparation of a sequence of three public lectures on the topic of engineering and design to be delivered at Princeton University. The text incorporates the subject of those Louis Clark Vanuxem Lectures, given in Princeton on December 7, 8, and 9, 2004, but is in no way a manuscript of the talks, whose titles were:

1. From Plato's Cave to PowerPoint: An Illustrated Lecture on the Illustrated Lecture
2. Good, Better, Better: The Evolution of Imperfect Things
3. The Historical Future: The Persistence of Failure

The written format has allowed me to expand the range of things and systems covered and to include more examples and detail than did the spoken word. Unfortunately, space in a book does not allow nearly as many illustrations as were used in my PowerPoint presentations at Princeton.

Engineers approach the lecture format in quite a different manner from that of humanists. In my experience, the latter typically read verbatim from a prepared text and use few, if any, graphics or illustrations. In contrast, engineers tend to use a good number of slides and related visual aids—in the form of drawings, diagrams, charts, graphs, equations, and demonstrations—to illustrate their talks, which are typically delivered extemporaneously. That is not to say that they are unprepared, for the engineer will more likely than not have gone over and over the visual materials and the essence of the commentary that will accompany them. The number and order of slides will be edited and reedited in the weeks leading up to the talk, for which the illustrations serve also much the same purpose as do prompting notes on index cards. Over the years, mechanical, visual, and digital devices ranging from magic lanterns to computer software like PowerPoint have greatly facilitated the process of preparing and projecting slides. Still, there remains room for improvement, as is described in the first chapter of this book.

Writing also benefits greatly from the use of computers, of course, but no author should ever blame malfunctioning electrons for the misfire of neurons in his own brain. If I have made any errors in this book, they are my responsibility alone and not that of the individuals who have helped me in so many ways. As always, I am deeply indebted to libraries and librarians, most importantly those at Duke University, and in particular Eric Smith and Linda Martinez. I am especially grateful to them for their assistance in helping me identify and secure obscure sources from incomplete references, and for introducing me to increasingly powerful electronic databases. And I am once again greatly in debt to the largely anonymous but immensely generous institution of Interlibrary Loan.

I am also grateful to Jack Judson, director of the Magic Lantern Castle Museum in San Antonio, Texas, who guided me through his outstanding collection of lanterns and related materials; to Tom Hope, who provided me hard historical data on the development of the slide projector; to Robin Young, who invited me and my wife to visit Stonecrop Gardens and who ensured that we had a good view of its flint bridge; and to Pete Lewis, who provided insight into and documentation for cast-iron bridges. Charles Siple, an inveterate correspondent and draftsman, was kind enough to draw the diagrams of splitting wedges and an arch from my amateur sketches. As usual, my family was also extremely helpful. Stephen Petroski helped me find documentation for my statements about design in sports, and Karen Petroski improved my knowledge of the Internet. Once again, Catherine Petroski was my first reader, and she also served as photographer and provider of digital images and graphics.

I was asked to write this book by Sam Elworthy of Princeton University Press. I am grateful to him for his persistence in convincing me to present the series of lectures and to prepare a book on the topic of design. The Princeton University Committee on Public Lectures and its chair, Sergio Verdú, extended to me the invitation to speak in the Vanuxem lecture series, which dates from 1912. It is an honor for me to join the distinguished list of Vanuxem lecturers.

Finally, I appreciate the planning, warm reception, and hospitality that members of the Princeton community extended to Catherine and me over the three days of the lectures. Susan Jennings and an excellent audiovisual crew made sure that the mechanical and electronic details were in order in the lecture room in McCosh Hall. David Billington, who was enormously generous with his time, turned me loose in his Maillart

Archive and allowed me to sit in on some of his own lectures and to meet with his students. David and Phyllis Billington were most gracious hosts, who helped Catherine and me experience Princeton in times of leisure and in time of emergency.

SUCCESS THROUGH FAILURE

INTRODUCTION

Desire, not necessity, is the mother of invention. New things and the ideas for things come from our dissatisfaction with what there is and from the want of a satisfactory thing for doing what we want done. More precisely, the development of new artifacts and new technologies follows from the failure of existing ones to perform as promised or as well as can be hoped for or imagined. Frustration and disappointment associated with the use of a tool or the performance of a system puts a challenge on the table: Improve the thing. Sometimes, as when a part breaks in two, the focal point for the improvement is obvious. Other times, such as when a complex system runs disappointingly slowly, the way to speed it up may be far from clear. In all cases, however, the beginnings of a solution lay in isolating the cause of the failure and in focusing on how to avoid, obviate, remove, or circumvent it. Inventors, engineers, designers, and common users take up such problems all the time.

The earliest useful things were, of course, those found in nature. Not surprisingly, these same things became the earliest tools. Thus, rocks came to be used as hammers. Whether a particular rock makes a good hammer depends on its size and shape and on its hardness and toughness relative to the object being hammered. Rock types that failed to accomplish desired ends became known as poor hammers and so came to be passed over. Better hammers resulted from eliminating the failures. However, even the best of rocks have limitations as hammers, and the recognition of their failure in this regard defined the design problem: Devise a better hammer. Among the problems with a hammer-rock can be that it is awkward or uncomfortable to wield. An improvement might be sought in the shape of the rock or in providing a handle for it—or from replacing the rock with something better. In time, a growing variety of metal hammer heads and wooden hammer handles, appropriate to a variety of tasks and grips, would reflect increasing specialization and diversification. Among such diversity, one might expect that there was a single best hammer for a particular task. All the others would fail to work as well at that task. Should all existing hammers fail to work properly for a newly developed task, then a still newer hammer might have to be developed. By the latter part of the nineteenth century, some five hundred different types of hammers were being produced in Birmingham, England, alone.

Technological systems also have their roots in the given world. The circadian and seasonal rhythms of nature drove the development of patterns of rest and migration. Even the simple act of sleeping when it is dark could be fraught with danger, however, as may have been discovered the hard way. If all the members of a group slept simultaneously, some might fail to survive the night. Recognizing this failure of the system would naturally lead to such concepts as the staggered watch and other means of protection. Thus, the group might begin sleeping in a cave whose single entrance could be guarded by a boulder rolled across it. The inconveniences of migration ultimately led to the development of systems of agriculture and defense. No matter how well developed a thing or system becomes, however, it will never be without limitations. There are no mechanical utopias. Therefore, there will always be room for improvement. The most successful improvements ultimately are those that focus on the limitations—on the failures.

Success and failure in design are intertwined. Though a focus on failure can lead to success, too great a reliance on successful precedents can lead to failure. Success is not simply the absence of failure; it also masks potential modes of failure. Emulating success may be efficacious in the short term, but such behavior invariably and surprisingly leads to failure itself. Thus, a single type of rock that worked reasonably well as a hammer for every previously known task might be said to be *the* hammer-rock. Whenever anyone wanted an all-purpose hammer they would look for that type of rock, if they had not already become accustomed to carrying it around with them. In time, however, there would arise a task in which the hammer-rock would fail. This would occur, for example, when the implement was used to strike a newly discovered but harder and tougher rock, with the purpose of shattering it. But to everyone's surprise, it would be the hammer-rock itself that would shatter. Past successes, no matter how numerous and universal, are no guarantee of future performance in a new context.

This book explores the interplay between success and failure in design and, in particular, describes the important role played by reaction to and anticipation of failure in achieving success. Since the book grew out of a series of lectures, the nature of lectures generally—and specifically the technology of the illustrated lecture as an evolving system—suggested itself as a topic with which to begin. From its precursors to the magic lantern, through the overhead and 35-mm-slide projector, to computer-based PowerPoint presentations projected through digital devices, successive improvements are shown to have been motivated by and arisen in response to the real and perceived failures of earlier means—and the systems within which they operated—to perform as well as could be imagined in the context of always-developing technologies and their attendant introduction of new expectations.

The vast majority of users of a technology adapt to its limitations. In fact, to use any single thing is implicitly to accept its limitations. But it is in human nature to want to use things beyond their intended range. Though a wooden pointer can be made only so long before it becomes too heavy and unwieldy to use on a stage, we invariably want to extend its reach. As a result, a lecturer might have to step into the field of a projected image to tap on a detail, thereby covering up some of the context. Of course, the limitations of the wooden pointer became moot with the development of the laser pointer, which has its own limitations. Its longer "reach" means that in an unsteady hand the movement of its "point" is amplified. Furthermore, the laser pointer's light touch does not allow the projection screen to be tapped for emphasis. Also, sometimes it is difficult to pick out the pointer's red dot from among a scattering of red data points. Technological advancement is not unqualified technological improvement.

Most things have more than a single purpose, which obviously complicates how they must be designed and how they therefore can fail. The more complicated the design problem, naturally the more difficult the solution and hence the more likely that some details and features may be overlooked, only to have their absence come to the fore after the thing is manufactured or built and put to the test of use. And failure can manifest itself in extrafunctional ways, including the inability of a product to maintain market share, thus disappointing corporate managers, directors, and stockholders. Poor performance, whether in the lab or on the ledger, signals a failure to be addressed. Such matters are explored through the many examples contained in the second and subsequent chapters of this book.

It is not only concrete things like projectors and pointers that pose problems in design and its limitations. Among the intangible things considered in the third chapter are intellectual and symbolic ones like national constitutions and flags, where the failure to anticipate how such politically charged things might not please their varied intended constituencies can be disastrous. Strategies for playing games like basketball, while perhaps of lesser consequence than political contests, are also matters of design, and the failure of a coach to defend against a boring offense or to match a hot shooter with a tenacious defender can result in a disappointing game for players and spectators alike. Successful design, whether of solid or intangible things, rests on anticipating how failure can or might occur.

Failure is thus a unifying principle in the design of things large and small, hard and soft, real and imagined. The fourth chapter emphasizes the sameness of the design problem for all sorts of things. Whatever is being designed, success is achieved by properly anticipating and obviating failure. Since earlier chapters focus primarily on smaller, well-defined things and contexts, this chapter also employs examples of larger things and systems, such as the steam engine and the railroad. With the underlying sameness of the design process established, the discussion turns to differences in the behavior of small and large things. In particular, the testing process, by which an unanticipated mode of failure is often first uncovered, must necessarily vary. Small things, which typically are mass produced in staggeringly large numbers, can be tested by sampling. However, very large things, which are essentially custom or uniquely built, do not present that same opportunity. And, because of their scale, the failure of large structures or machines can be devastating in all sorts of ways, not the least of which is economic.

The remaining chapters focus exclusively on large things. The fifth chapter considers buildings, especially tall and supertall buildings. Though the desire to build tall did not originate with the skyscraper, it is in that genre of architecture and structural engineering that failure can have the most far-reaching consequences. The decision to build tall is often one of ego and hubris, qualities that not infrequently originate in and degenerate into human failings of character that can lead to structural ones. In the twenty-first century, limitations on the height of buildings are not so much structural as mechanical, economic, and psychological. Structural engineers know how to build buildings much taller that those now in existence, but they also understand that height comes only at a premium in space and money. The taller buildings go, the more people must be transported vertically in elevators. The more elevators that are needed, the more elevator shafts must be provided, thus taking up more and more volume. This reduces the available office space per floor, which in turn threatens the economic viability of the enterprise. Nevertheless, for reasons of pride and striving, taller buildings will continue to be built. Still, no matter how many supertall buildings stand around the world, their success does not guarantee that of their imitators. The collapse of the twin towers of the New York World Trade Center demonstrated that unanticipated outside agents (and unperceived internal weaknesses) can create scenarios that can trigger novel failure modes.

In the sixth chapter, the book's focus turns to bridges, which provide a paradigmatic study of the paradoxical nature of success and failure in design. Overconfidently building increasingly longer bridges modeled on successful prior designs is a prescription for failure, as has been demonstrated and documented repeatedly over the past century and a half. The designers of the first Quebec Bridge, for example, were emboldened by the success of the Forth Bridge and set out to better it with a lighter and longer structure of its type. Unfortunately, the Quebec collapsed while under construction, an event that gave the cantilever form upon which it was based a reputation from which it has yet to recover in the world of long-span bridge building. Though the Quebec Bridge was successfully redesigned and rebuilt and stands today as a symbol of Canada's resolve, no cantilever bridge of greater span has been attempted since. The Tacoma Narrows Bridge, the third longest suspension bridge when completed in 1940, proved to have too narrow and shallow a deck, which accounted for its collapse just months after it was opened to traffic. A relatively unknown engineer without his ego invested in the design had actually warned against the excessive narrowness of the deck, but his objections were overcome by the hubris and influence of the design consultant, whose confidence in his theory was backed by numerous prior successes. Such examples provide caveats against success-based extrapolation in design. Past success is no guarantee against future failure.

The final chapter looks at the historical record of colossal failures, especially in the context of the space shuttle program and of long-span bridges. In the case of bridges, there is a striking temporal pattern of a major failure occurring approximately every thirty years since the middle of the nineteenth century and continuing through the millennium. All of the half-dozen remarkable failures that occurred within this time span resulted from designs based on successful precedents rather than on a more fundamentally circumspect anticipation and obviation of failure. Such compelling evidence argues for a greater awareness among designers of the history of the technology within which they work, but such looking back is not generally in the nature of forward-looking engineers working on the cutting edge. Still, the historic pattern has been persistent, and it should be convincing. It even suggests that a major bridge collapse can be expected to occur around the year 2030. Such a prediction gains credibility from the fact that bridge building in the twenty-first century continues to go forward in a way not unlike that which preceded the failures of the Quebec, Tacoma Narrows, and other overly daring bridges. But failures are not inevitable, of course, for if they were there might be no technological advancement. Indeed, future failures can be anticipated and thereby avoided through an appreciation for the past, which reveals in case after case an incontrovertible if paradoxical relationship between success and failure in the design process generally.

Failure and responses to it may not explain every aspect of

every design, but from the engineering perspective of this book it is presented as a unifying theme for describing the functional evolution of things. In particular, the interplay between failure and success in the development of technological artifacts and systems is presented here as an important driving force in the inventive process. Most of the examples are drawn from the fields of mechanical and civil engineering, in which the author has the most direct experience. There are, of course, countless examples besides mechanical devices and civil structures that the reader may call to mind to test further the paradoxically opposed hypotheses that failure drives successful design and that success can ultimately threaten it. But the genesis of this book dictated that it not take up more than a narrow space on the library shelf, and so it could not be overly wide ranging. Hence its focus on the functional. There are numerous other factors that affect design—including the aesthetic, cultural, economic, egotistical, ethical, historical, political, and psychological—but no single book can hope to say everything about everything.

1

FROM PLATO'S CAVE TO POWERPOINT

Should we not illustrate our lectures,
and cease to lecture about our illustrations?

—C. H. Townsend[1]

Imagine anyplace, anytime. There, on a cloudless night, the shadows cast by a full moon light the landscape as if it were a stage. A quiet and patient observer could watch the limned images of wildlife morph out of the wings and stalk and flee in pantomime, playing out a nightly drama under the stars. As the moon progresses across the sky, the shadows on the ground slowly but inexorably follow their own circuit, shortening and lengthening around the moondial. At dawn, the footlight of the sun signals a new act.

From the beginning of the solar system, the stage has been set for such dramas in light and shadow. Given that, they are unremarkable. Yet the unremarkable often serves as the basis for the remarkable. Any light will, of course, cast shadows. Fire, being a flickering source, can introduce an added eeriness and nervousness to the camp theater in the round, the shadow players thrown back as it were from the heat. For millennia, our ancestors passed nights by the light of flames keeping an audience of wildlife at bay.

Daylight produced an inversion of the set—and views we have all on occasion seen were created "entirely independent of man's invention and control." In a room darkened by shutters against a bright sun, the wall opposite can display a sharp inverted image of an outdoor scene. A distant grazing cow, perhaps, or a floating cloud, can be carried on light squeezed through a hole in the woodwork like water through a small breach in a tank. Likewise, on the shaded ground beneath a tree, an image of the sun can be thrown down through "chinks between the leaves." These are perfectly natural phenomena that take no lens or sleight of hand to produce.[2]

More often than not, however, what we see projected is the product of design. Plato's allegory of the cave makes much of a controlled drama of illumination and images. In the dialogue with his young follower Glaucon, Socrates describes prisoners who have lived since their childhood in a den, where they sit with their backs to the entrance and are restrained so that they can look only at the wall directly in front of them. Behind them, fires burn and provide a source of light. Between the prisoners and the flames there is a raised walkway, and anything that moves across it casts a shadow on the wall. To the prisoners, these shadows are the extent of their experience and so become their reality.[3]

After describing the situation in the cave, Socrates posits that one of the prisoners be set free and be allowed to turn toward the mouth of the cave and see the fires and also the players and their burdens that have been casting the shadows. Which will be more real to the freed prisoner, Socrates asks, what he now sees in the flesh or what he has seen as shadows all his life? And if the prisoner is forced to look directly into the source of light, Socrates asks further, will he not be blinded and wish to turn back toward the wall of the cave, where images are sharper and more familiar?

Socrates then imagines that the prisoner is taken out of the cave and exposed to the direct experience of the sun and everything that it illuminates. At first the prisoner would be blinded by the brightness, but in time he would come around to see the world outside the cave for what it is. If he then returned to the cave and sat among the prisoners who had remained there, his descriptions of the sources of the shadows and external reality would be met with skepticism. Better to remain in the cave, the prisoners would say, rather than to go away and come back without clarity of vision.

Great philosophical and technical advances have been made since ancient times, with varying contributions to understanding reality and capturing it in media more tangible than shadows. The camera obscura enabled artists to capture fleeting glimpses of reality undistorted and in proper perspective, albeit upside-down. David Hockney has argued that Renaissance painters employed such technology to produce their almost "photographic" masterpieces.[4]

The use of chemical fixing to freeze images in the optical camera—which photography pioneer William Henry Fox Talbot referred to as "the pencil of nature"—made it possible to stop the wings of birds in flight, the tails of cats in free fall, and the hoofs of horses in full gallop. This technological development and its consequent mechanical realism paved the way for nonrepresentational modern art. Now, digital imaging has made it possible to graft the head of a lion onto the body of an eagle, providing evidence for the existence of griffins, if we can believe our eyes.

Developments in optics, chemistry, electricity, and computers have at the same time freed us from Plato's cave and shackled us within another. The allegory of the cave updated to modern times might be set as follows. A group of people is seated in a cavernous room, restrained by a prevailing paradigm. The chairs in which they sit are rigidly attached to the floor and to each other, and images on the screen before them rivet the group's eyes to it. They are watching things that are being projected from a booth in the back of the room, which they sometimes forget they are in. On occasion, a head casts a shadow on the screen, and members of the group move slightly to check if it is theirs. The images on the screen are accompanied by commentary coming from a disembodied voice issuing from speakers distributed around the room. Now and then a red dot moves about the images, like a fly about a horse, and lights upon a point. The voice continues to describe the images and read the words projected on the screen. The images and words are sharp and bright and are the reality of the moment. They fade in and fade out like shadows on a night of patchy clouds.

This modern Plato's cave could obviously be an auditorium in which a PowerPoint presentation is being made. PowerPoint is, of course, the computer program produced and marketed by Microsoft, but it is also a thing, in the sense that it was invented and has been designed, and redesigned—by maker and user alike. It continues to evolve with ever more bells and whistles and ever more clever new uses. However, unlike a bottle cap or an umbrella, this thing is not something we can grasp in our fingers or hold in our hand. It is intangible. It is not hardware. It is software, designed for use within a system of computers and projectors and screens and speaker and audience. PowerPoint is a thing that enables its user to accomplish an end, namely, the design of a "slide show," which is also a thing of sorts. Thus the language of PowerPoint, like the language of everything made, harks back to some previous thing, for long before there were digital computers, there was a need to communicate visual images to an audience in such a manner that the images could be viewed simultaneously by everyone present.

Among the oldest surviving permanent works of art are drawings on the walls of caves in France, India, and elsewhere around the world. These cave drawings may have been made not strictly for aesthetic reasons but as illustrations and schematic diagrams before which gathered novices could be instructed by elders in the art of war, or before which hunting parties could be briefed on strategy prior to departing on a sortie. It has been speculated that ancient petroglyphs found in California were drawn to record earthquake activity in the area.[5] Perhaps these primitive rock paintings also served to assist in communicating the nature of and the responses to earthquakes.

Stone carvings and inscriptions are as old as civilizations. Among surviving hieroglyphics are depictions of how heavy statues were moved, diagrams of obvious instructional value for a team of haulers assembled before them. Obelisks have been inscribed with diagrams showing how they were erected. Likewise, inconspicuous stones and timbers found in vaults, attics, and other less accessible places in Gothic and medieval structures have revealed the sketches and calculations of stonemasons and carpenters—perhaps merely scratched out to clarify their own thinking or drawn for the instruction of an apprentice. To this day, it is not uncommon to find such essaying or instructional graffiti on furniture or at building sites.

Venues for formal and informal education have long been places where visual aids have been employed. As recently as half a century ago, blackboards were standard equipment in classrooms and lecture halls, and many teachers and professors prided themselves in their board work. But blackboards, like all made things, had their limitations, among which were the propensity to get dusty after too many erasings and the difficulty of being read in less than ideal light. So-called white boards, which were often touted as successors to blackboards, became common in the latter part of the twentieth century. The multicolored pens that were used with white boards were ostensibly a great improvement over chalk, which is notoriously brittle. Unfortunately, white-board pens emit distracting if not intoxicating fumes and tend to dry out, leaving a mark too faint to be easily seen.

Not every lecturer was able to develop a facility with writing and drawing on a black- or a white board. Though some prided themselves in their (usually well-practiced) ability to produce Palmer-quality penmanship on the vertical surface of a board, most could not even keep their lines level or their inclines parallel. Drawing was especially difficult for many a nonartist, which was a considerable liability for a naturalist or architect attempting to construct an accurate image of flora, fauna, or facades. Hence, any device that could be employed to project drawings made carefully at one's leisure, or pictures cribbed from nature or from sources with access to better draftsmen or, at later times, photographers, would have been

enthusiastically welcomed. However, according to one historical survey of optic projection,

> No one knows who first designedly arranged a darkened
> room with a white wall or screen on one side, and on
> the other a small opening facing some object or scene that
> could be brightly illuminated. All we know is that the ear-
> liest accounts of the pictures in a dark place are in connec-
> tion with the explanation of some other phenomenon, and
> not to show that such pictures were possible. It was also
> recognized in the first statements, as in the works of Aris-
> totle and of Euclid, that [just] as light rays extend in
> straight lines, . . . those from an object must cross in pass-
> ing through a small hole, and hence the images beyond
> the hole in the dark place must be inverted, the top being
> below and the right being left.[6]

The problem of projection was thus not new even in ancient
times, and neither were solutions in the Middle Ages. The
camera obscura, or "dark chamber," though not necessarily by
that name, was described as early as the eleventh century[7] and
was mentioned by Leonardo in the late fifteenth.[8] It was embodied in the principle of a pinhole in one wall being the
source of an inverted image projected upon the wall opposite.
Using an artificial light source in place of the sun would have
enabled images to be cast upon a wall even at night. As early as
the fifteenth or sixteenth century, Sicilian priests were reported
to be "using lanterns of undescribed construction with hand-
painted slides" to produce visions.[9] Since visions tend to be
ethereal anyway, the quality of the projected image need not
have been very sharp to be effective. Giambattista della Porta,

whose *Magia Naturalis* appeared in the mid-sixteenth century, first described using "a convex lens to perfect the images and of placing transparent drawings opposite the opening." Furthermore, "To these drawings he attached movable parts, and thus produced astonishing effects, which the unlearned ascribed to magic, a term connected with the lantern ever since."[10] (In the mid-nineteenth century, after photographic slides began being used in the "magic lantern," one lecturer would wish "that some more scientific, if not so familiar, a name for our instrument were recognised.")[11]

In the seventeenth century, improved versions of the *laterna magica* were developed and used by physicians, mathematicians, and natural philosophers, including Johannes Kepler and Christiaan Huygens, who is often incorrectly credited with the invention of the magic lantern in the 1650s.[12] Naturally, the light source was a critical component of the projector, and "an elegant version" incorporating "a polished tin cylinder holding a concave mirror and candle" was made in 1671 by Athanasius Kircher. During the eighteenth century, the magic lantern—also to be known as a sciopticon or stereopticon—was used to amuse children. Later, entrepreneurs "employed the device to produce optical illusions for the deception of spiritualistic groups." These illusions were principally phantasmagoria,[13] a term that came to mean an elaborate display of optical illusions and effects by which "terrific figures are produced, which seem to approach the audience from an amazing distance and then recede again; or they rise to the ceiling apparently, and then descend to the floor."[14] Such uses of a hidden hand-held version may have earned the magic lantern the label "the lantern of fear."[15]

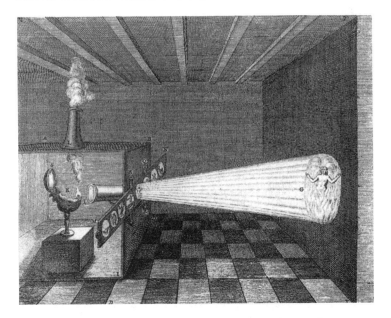

An early magic lantern was illustrated in the second edition of
Athanasius Kircher's treatise on light and shadow. (From *Ars Magna
Lucis et Umbrae*, 1671.)

In the nineteenth century, the basic magic lantern was typically made of tin, had an oil lamp inside, and was fitted with a
chimney through which the smoke was carried away. A lens or
set of lenses, known as "bulls' eyes," were placed in front of the
source of light to diffuse it uniformly as well as to project
whatever object or image was placed in its path.[16] Though oil
lamps may have been adequate sources of illumination for
seances or small shows, better ones were desirable for scientific
lectures and large extravaganzas. Early in the nineteenth century, the American chemist Robert Hare "discovered that a

flame of oxygen and hydrogen blown against lime rendered it incandescent, given a dazzling light," which came to be referred to as limelight. English inventors adapted and developed the phenomenon to the point where Henry Langdon Childe "projected his pictures on huge screens in the largest halls of London." His use of "dissolving views" in conjunction with limelight, which he would employ in his "Grand Phantasmagoria" on the occasion of the 1838 opening of the London Polytechnic Institute, "became an essential part of lantern projection" from the mid-1820s through the end of the century.[17] By the early twentieth century, smaller magic lanterns were "much in vogue for Christmas presents."[18]

Although popularly developed as an entertainment medium, by which patrons could watch a picture show for a fee, ultimately magic lanterns and lantern slides "had the greatest impact on educational lectures, especially in visual disciplines. They played a vital role in the development of disciplines such as art and architectural history, making possible the detailed study of objects and sites from around the world."[19] By the end of the nineteenth century, "the application of lantern slides for educational purposes was realized, representing the first audiovisual format used in an era of increasing interest in visual education."[20] Regardless of their application, throughout this period various new and improved sources of illumination for lantern slides were developed, including kerosene lamps employing flat wicks, lamps burning magnesium, and carbon arc lamps, which became more convenient as electric power stations obviated the need for a large bank of batteries. Eventually, sufficiently powerful incandescent bulbs were developed and used where a source of electricity was available.[21] In New

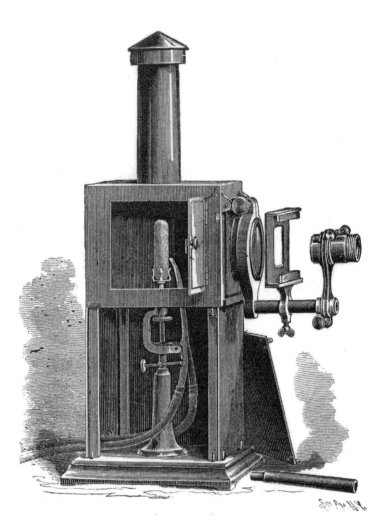

Limelight was still employed as a source of illumination in some early
twentieth-century "scientific lanterns." (From *Scientific American
Supplement*, 1905.)

York City in the early twentieth century, there were art lecturers "showing colored photographs of the great paintings of Europe, who have never seen the originals."[22] Lantern slide shows "were a substitute for expensive and physically difficult travels" and also "rivaled the dime novel as a source of adventure."[23] One critic of the form could write in 1912, "Lecturing with lantern illustrations has so nearly superseded the well-prepared, authoritative discourse, that the latter has become a rarity."[24] In the late nineteenth and early twentieth centuries, the term "illustrated lectures" was synonymous with "public lectures illustrated by magic lantern slides."[25]

As bright as the illumination source may have been, an illustrated lecture was only as good as its images. The oldest slides were naturally hand-drawn. In the 1840s, shortly after the introduction of photography, daguerreotypists looked to the lantern as a means of projecting photographic images. However, since daguerreotypes were opaque, another medium had to be employed. The brothers William and Frederick Langenheim, of Philadelphia, adapted a French process to produce glass negatives, from which they printed positive images onto glass slides suitable for projection. Their first photographic lantern slides were produced and shown by the Langenheims in 1849, and "by the summer of 1851 they had published 126 slides, including views of Philadelphia, Washington, and New York, as well as portraits of prominent personalities of the time." The brothers have been credited with having inventing the "slide photograph," which was an inexpensive alternative to hand-drawn slides. They patented their "hyalotypes,"[26] which used glass for both negative and positive impressions,[27] and displayed them at London's Great Exhibition. The Langenheim firm came to be "well known as the best manufacturer of glass slides for the magic-lantern," and in promoting the slides used "as many as eight magic-lanterns or stereopticons directed at the same instant upon the same screen," no doubt using a variety of effects.[28]

The success of the Langenheims naturally attracted many competitors. In 1874 the Philadelphia firm of Benerman & Wilson was offering a wide selection of views, including exclusively a "large stock of the celebrated productions of Messrs. J. Levy & Co., of Paris." According to an advertisement that appeared in *The Magic Lantern*, the slides included views from all over Continental Europe, the Far East, and America. "Plain, colored, and comic lantern slides of all kinds" could be had at Benerman & Wilson "at low prices." A "lantern outfit" could be had for one hundred dollars.[29] The firm also supplied sciopticons, which were "superior and popular" magic lanterns invented and made in Philadelphia by Lorenzo Marcy. The sciopticon was capable of providing "a very powerful light, giving a brilliant, 10-foot-diameter disc" and had "larger than usual" condensing lenses.[30]

In an early commentary on the use of lantern slides, projected to the "extraordinary dimensions of 20 feet in diameter," before "a large number of spectators at one time," the application was hailed to be "as interesting and as beautiful as anything connected with photography." The "perfect photographs" were "placed behind the condensing lenses of the oxyhydrogen lantern," and the slide show was described as "the exhibition of what are technically termed 'Illuminated Photographs.'" It was judged "best to use two lanterns and show the photographs as 'dissolving views.'" But, when showing slides of sculpture, "a much more effective result" was believed to be "obtained by first throwing upon the screen, with the spare lantern, a disk of blue light."[31] Action slides mounted in frames fitted with levers, cranks, or other means for alternating between scenes were used to produce moving images. By flipping between two or more related views, animated cartoons (such as a man tipping his hat)[32] could be produced. The English slide painter C. Constant "made himself immortal by painting the original of the world-famous slide of the sleeping man swallowing rats."[33] The action was often accompanied by "spluttering and chomping—with all of the essential and undignified sound effects being provided by the lecturer or lanternist."[34] The "gentleman in bed who swallowed the rats" could be the "high spot of the whole entertainment." As one attendee of a "Magic Lantern show given to poor and destitute children" recalled the gentleman years later, "He was our star turn, our living picture, our cinema. He was everything. He made up for all the long, and sometimes dreary lectures to which we had to listen, even though it was 'accompanied by Dissolving Views.' We wanted life and movement, and when he was thrown on the screen the grand climax of the whole entertainment had been reached, and enjoyment was complete."[35]

As much as audiences enjoyed all or part of illustrated lectures, there were—as there always are—problems with the technology. Early glass photographic slides were mounted in wood frames that were far from uniform in size, and those from one manufacturer did not necessarily fit into the lantern of another. One British critic wrote that "we ought to adopt a standard gauge for our glasses, say three-and-a-half inches square for views for the general run of lantern."[36] When they did become more or less standardized, a typical European lantern slide measured 3¼ inches square, while an American one measured 3¼ × 4 inches, its oblong shape making it easier to orient.[37] Regardless of its size, the slide was typically assembled out of two pieces of glass. On one was printed the photographic image, which was protected by placing the second piece of (plain) glass over it. Paper tape around the edges not only held the parts of the slide together but also prevented debris from getting in between the glasses. The picture sandwich usually incorporated an opaque paper mask, mat, or mount that "served to exclude extraneous details at the edge of the picture, and mounts of circular, oval and other shapes were considered an aesthetic enhancement," which could also be imprinted with the name of the manufacturing firm. The paper insert "also served to mask the shortcomings of cheaper lenses, which tended to soft focus towards the edges of the picture."[38]

Though hand-painted lantern slides in color were long common, the color-photographic slide began to appear only in the twentieth century, "utilizing some of the earliest color photographic processes."[39] In 1929 the Chicago Public Library maintained a collection of fifty thousand lantern slides—black-and-white and color—for circulation.[40] Smaller selections were often stored and transported in cases about the size of a shoebox and fitted with a carrying handle. Since keeping slides in proper order and orientation was always a concern, some boxes were designed to open flat into two equal halves. As the slides, which had to be loaded one at a time, were shown, they could be put in order (but upside-down) into the empty top half of the box. When the show was over and the box closed, all the slides were back in order.[41] The motion picture

Mona Lisa Smile, in which the protagonist is an art history teacher at Wellesley College during the 1953–54 academic year, when lantern slides were still in use in her discipline, makes clear the lasting importance of slide boxes.

Outside the academy and learned society, the popularity of lantern-slide shows had long before then begun to decline. By the early 1890s, Thomas Edison had developed a system of taking motion pictures on celluloid film, and the brothers Louise and Auguste Lumiere gave their first public showing of their cinematographique in 1895.[42] Early cinematic apparatus was often "combined with a slide projection facility," so that advertisement and coming attraction slides could be shown while film reels were changed.[43] According to one inventor, "the moving picture machine is simply a modified stereopticon or lantern, i.e., a lantern equipped with a mechanical slide changer."[44] Whatever its roots, the rapid rise of motion picture technology stole the show. The lantern slide never regained its stand-alone place in mass entertainment, but it remained popular into the early 1960s and continued to be used for presentations at large meetings, where it could be used to project images sixty feet wide.[45] As late as the mid-1970s, lantern slides were still being used in medical schools and hospitals, which tended to have large, specialized slide libraries.[46]

In the mid-1930s the Eastman Kodak Company introduced Kodachrome film, thus making possible the relatively inexpensive "color transparency" slide, which "could be produced from the newly popular 35-mm cameras." This medium provided a lighter, more compact (2×2 inch mount size), and relatively unbreakable alternative to the lantern slide for home use and for amateur lecturers, but the quality of the early 35 mm image was inferior to what professionals had become accustomed to in glass lantern slides. However, in the 1950s "color film emulsions had been improved to the point where a 35 mm image could be projected to a 60-foot width," thus making the 2×2 inch slide a full competitor.[47] Kodak introduced the slide tray in 1950 and offered a fully automatic Cavalcade projector in 1958.[48] (As early as the late 1880s, magic lanterns fitted with clockwork mechanisms had been used to advance advertising slides automatically, and motor drives were employed in the 1920s.)[49]

When first offered, color transparencies were "quite expensive, but their toughness and lightweight construction numbered the days of glass slide utility."[50] Still, some art historians insisted that the smaller slides simply did not provide as sharp an image as glass lantern slides, and so they, like the Wellesley instructor, continued to use the old technology. With the introduction of Kodachrome II film and the Kodak Carousel projector in 1961, however, the glass lantern slide was essentially made obsolete. (But no technology remains dominant forever. Though thirty-four companies offered slide projectors in 1979, the year of peak production, only seven did so in 2000.[51] In 2003, with the use of computer-generated and computer-projected slides well established, Kodak announced its plans to discontinue making slide projectors in 2004 and to cease servicing and supporting them after 2011.)[52]

Not all lectures or presentations required high-quality photographic slides. Especially for situations that did not rely so much on pictures as on drawings and diagrams, the development of the overhead projector provided a quick and easy alternative to slides of any kind. A bulky magic-lantern model employing a

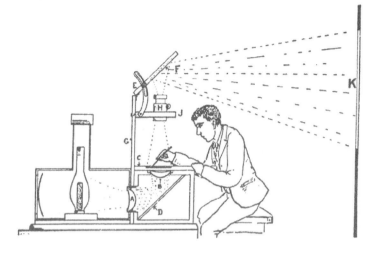

A magic lantern could be configured to serve as an overheard projector. (From *Boy's Own Paper*, 1900.)

periscope-like principle was illustrated as early as 1900 by the English "optical experimenter" and journalist Theodore Brown as a means of "drawing live cartoons on the screen."[53] More compact overhead projectors date from the mid-1940s,[54] when they were developed as an aid in police work. They soon found their way into bowling alleys and schools, where they projected scores and lessons as they were written down in real time. Some overhead projectors were fitted with a reel of clear plastic film that was stretched over the light table and advanced by the rotation of a crank attached to a take-up reel on the other side of the projector's base. When a section of the film had served its purpose, the film was advanced to position a clean section over the light table. Soon, even the most tradition-bound scientists and engineers, who eschewed slides as a medium too rigidly set for their extemporaneous presentations, began to embrace the use of the overhead projector, using $8\frac{1}{2} \times 11$ inch transparencies. After all, it was not so radically different from writing on the blackboard, and it allowed the writing to be done on a more natural horizontal surface and permitted the use of colored pens to clarify points.

Handwritten notes and hand-drawn charts were generally considered too informal to project at serious business meetings, where attendees dressed up in suits and ties. So the overhead projector did not come into widespread use for serious business presentations until the mid-1970s, when finished-looking transparencies (also known as "overheads")[55] could be made by photocopying professionally typed and drawn material.[56] Now, anything that could be photocopied could be projected onto a screen, including photographs, pages of books, computer printouts, and charts and graphs. Transparencies of these latter were especially popular among scientists and engineers, who could overlay one atop another to compare different sets of data. Some technical presentations were prepared by taping multiple overlays together to form perfectly aligned transparent flip charts.

The overhead projector was not without its limitations, of course. It was bulky and usually sat on an even bulkier table or cart directly in front of the screen. The presenter had to stand beside it to access the stack of transparencies that constituted the presentation. They were usually placed on the table to one side of the overhead projector, with a space on the other side of it being reserved for piling up the used transparencies. The projector itself and the presenter standing beside it blocked the view from what might have been considered the most desirable seats in the room. (Presenters had been known to speak while seated beside the projector, which did improve visibility of the screen for the audience, but such a practice was generally frowned upon as poor etiquette in contexts where presentations were customarily made from a standing position.)[57] To alleviate this problem somewhat, the image was often projected high on the screen, which in turn created the problem of distortion known as "keystoning," in which the top of the image was noticeably wider than the bottom.[58]

Like most modern projectors, the overhead was noisy and thus required the presenter to speak up. The cooling fan that produced the noise typically blew air out the side of the projector, which often was exactly where the pile of transparencies had been placed before the fan was turned on—or where they were placed after being shown. This could and often did lead to overheads sailing off the table and getting all mixed up. (The savvy presenter numbered the set of transparencies and kept them in order in a box or in plastic sleeves in a ring binder.) Since the presenter's focus was naturally on properly orienting and centering the next transparency on the overhead, used transparencies were often tossed down carelessly and into disarray. If, after the talk, a question were asked, say, about the second graph shown, considerable shuffling and confusion was likely to ensue. Overall, the overhead projector had a considerable number of failings, but because a set of transparencies could be created at the photocopier in the minutes before a meeting or lecture, it remained a popular choice over the alternative.

The alternative remained, of course, the use of 35 mm slides, which could be made simply by inserting frames of 35 mm film into 2×2 inch cardboard or plastic mounts.

However, to create such artifacts and thus a presentation, the images for the slides first had to be collected or prepared and then photographed under proper lighting conditions, and finally the film had to be developed before being cut into frames to be made into slides. Typically, a professional graphics department did most of the work, which required a considerable lead time. This began to change with the introduction of the computer, and by the mid-1970s computer graphics software made it possible to compose slides on a terminal's screen.[59] However, like much hard-copy computer output at the time, the slides still had to be picked up at a central computing facility, where the slide-making hardware resided.

There were other problems with slides and slide projectors. Like the overhead, the slide projector was typically placed directly in front of the screen. However, since it was not as bulky and could be fitted with a remote control, the presenter did not have to stand beside it, so the good seats were not preempted—as long as those sitting in them could hear the lecturer over the fan motor. Depending on the limitations of the available lens, the slide projector might have to be closer to or farther from the screen to get the desired image size. The start of many a presentation was delayed while an extension cord was being tracked down, or a book or other object of the right height was located to serve to prop the projector up at a suitable angle. This necessitated compromising between maximum image height and minimum keystoning.

In America in the latter part of the twentieth century, the most popular slide projector was the Kodak Carousel. Like any consumer product, it came in a variety of models (designs), and it evolved over the years by removing the little (and big) annoyances of operation that frustrated early users. These included bulbs burning out, slides melting, and mechanisms sticking. By the 1980s the machine had finally developed into a highly reliable device that was largely user friendly and trouble free. It might be said to have been "perfected."

That is not to say that making a 35 mm slide presentation was without risk. The safest way to insure that one's slides were all in the right order and in the correct orientation was to load them into a carousel well before the lecture, lock them into place, and run through them one last time. This could mean that one had to travel with a relatively bulky carousel and generally not be able to preview or practice the presentation without a projector. The presenter who forgot the order of the slides had to make adjustments to the lecture on the fly.

The alternative was to carry the slides in a pile secured with a rubber band or in an envelope or a box or arranged in transparent plastic sleeves. This enabled the presenter to review and shuffle the slides up to the last minute, when they typically had to be loaded into a carousel amidst the chatter before a meeting—a process that was fraught with danger. Loose slides obviously could get out of order, and even ordered slides could be improperly inserted into the carousel. Because of the optics of the projector, slides have to be inserted upside down for proper projection. (For rear-screen projection, they have to be inserted both upside down and backwards.) It can be a challenge to look at a 35 mm slide with the naked eye in a dimly lit auditorium full of people talking to tell what is up, down, front, back. (Wise presenters learned to put a distinguishing mark at the lower left front of the upright image, so that when the slides were properly inserted all the indicia showed above the notches in the carousel slots. Lantern slides had been keyed in much the same way.)

In spite of all the pitfalls, a well-prepared and well-executed slide presentation in a well-equipped auditorium can still be a thing of beauty. Preferably, the slide projector is enclosed in a projection booth in the back of the room, more or less at the same height as the large screen in the front, thus eliminating the problems of noise, keystoning, and head shadows. The speaker stands at a lectern beneath or to one side of the high screen and has a remote control that is either hardwired to or in infrared or wireless contact with the projector. The speaker is fitted with a lapel microphone, so that her voice is not lost when she turns to look at the screen. Were it not for the fact that 35 mm slides are inconvenient to create, store, carry, and arrange—though not nearly so as lantern slides—they might have remained the medium of choice for presentations of all kinds.

However, with the introduction of the personal computer in the early 1980s, the technical groundwork was laid for a new alternative that addressed the theretofore accepted short-comings of the 35 mm slide and its cumbersome infrastructure. Assigning absolute credit for the development of a computer application, like that for any invention, can be difficult, since ideas tend to be in the air and feed on one another. Someone may suggest the germ of a concept, but another person might carry it to fruition. The problem is not unique to computers. Samuel F. B. Morse engaged in long and acrimonious debate and legal wrangling with Charles Jackson, a physician-geologist from Boston who claimed that the idea for the telegraph was their "*mutual* discovery," made in 1832 while they were both crossing the Atlantic aboard the ship *Sully*.[60]

In 1981 Whitfield Diffie was a mathematician and cryptography expert who was working on problems relating to telephone system security at the Bell-Northern Research Laboratory. By his own account, in the course of preparing to put together a 35 mm slide presentation, Diffie wrote a computer program that could draw a frame on a sheet of paper. He subsequently augmented his program to draw a number of such frames on the same sheet and to allow text to be included inside them. The result was the layout for a "slide show," which could be given to a graphics department, which in turn could return to Diffie a finished set of 35 mm slides for his presentation. Arguably, this was the germ of the idea for developing computer software that could produce a slide show that would also run on the computer, thus eliminating the need for the physical slides. According to Diffie, it was a colleague, Bob Gaskins, who appreciated the value of such an application and who "was the one who had the vision to understand how important it was to the world."[61]

According to Ian Parker, who interviewed both Diffie and Gaskins for a *New Yorker* article on PowerPoint presentations, Gaskins observed others in the laboratory using ill-suited computer equipment to make overhead transparencies and got the idea to design a graphics program that could create and edit virtual slides. In 1984 he left Bell-Northern to join Forethought, a struggling Silicon Valley software company. Gaskins and Dennis Austin, a software developer, began working on a program they called Presenter, which to avoid a trademark problem was to be renamed PowerPoint before first going on sale in 1987. PowerPoint 1.0 was a black-and-white Macintosh application that enabled users to print out pages that could be photocopied onto overhead transparency film.

Soon, Microsoft acquired Forethought and with it Power-Point. The first version of PowerPoint for Windows came out in 1990, and the much-maligned AutoContent Wizard, which ostensibly helps the user create a presentation, was added in the mid-1990s. In the meantime, digital-image projectors became increasingly available in classrooms, meeting rooms, and auditoriums, thus enabling the contents of the computer screen to be displayed before a large audience. As if foretelling the imminent triumph of the newer technology over the old, such devices were referred to simply as "projectors."

In short order, the production of 35 mm film slides began to level off and decline. By the late 1990s PowerPoint was available in thirty-five languages and accounted for an increasing number of slides of all kinds, and as many as 70 percent of the nearly two billion "professional electronic slides" made worldwide. The growth continued, even though early electronic slides were "inferior to film slides in color and resolution." Among their greatest advantage was that the "slide is ready as soon as it is designed."[62] Now, of course, PowerPoint is as generic a product name as Kleenex and Xerox.

When PowerPoint was still a relative novelty, audiences were surprisingly patient with the "new technology," and this unelaborated-upon term was muttered frequently before, during, and after an introduction that could not be stretched out any further. The speaker having been introduced but the slide show not yet ready, everyone just stood around or sat by and watched in silence as the bashful new technology was coaxed out of its black box. As recently as the late 1990s, meetings could be stopped dead for twenty minutes while technicians, speakers, and well-intenders huddled over a recalcitrant laptop computer trying to project a PowerPoint presentation on an auditorium screen. Such incidents led many a PowerPoint presentation to be backed up with overhead transparencies or 35 mm slides (or both), which continued to be carried to meetings by many a presenter.[63] Now, laptops of all makes and models interface easily with projectors, and inserting a CD into a strange laptop and calling up its contents usually, but not always, goes off like clockwork. Nonetheless, some lecturers continue to rely on multiple media and redundancy to put their mind at ease.

If nothing else, PowerPoint presentations are generally neat and uniform in format. But the software that "was developed to give public speakers control over design decisions"[64] about the appearance of their slides is no substitute for the sense of proportion and restraint that professional graphics departments once provided, and that not every speaker has. Many of the "presentation designs" offered by PowerPoint to give unity to a presentation are cutesy at best. They should be, but surprisingly are not, out of place in a professional setting. Many of the "AutoLayouts" for text, bulleted text, tables, charts, and clip art are pat and almost juvenile. While PowerPoint does allow plenty of room to adjust and rearrange the set layouts, doing so risks disturbing the visual balance even more. Regardless of its pros and cons, becoming familiar with all the complex software that is PowerPoint can be daunting. Thus, it should not be surprising that workshops have been developed to provide the neophyte with "the only training you'll need for creating powerful

and persuasive presentations!" According to the brochure an-
nouncing one such workshop, "Today's high-tech audiences
are accustomed to high-powered multimedia extravaganzas.
They not only want to hear your message, they expect to be en-
tertained at the same time." By attending a one-day workshop,
you can "Learn how to captivate any audience with just the
right mix of multimedia magic!"[65] The language is remarkably
reminiscent of reviews of early magic lantern shows, and the
animated effects suggestive of phantasmagoria.

Most of the criticism of PowerPoint has been leveled not at
its superficial qualities but at the superficial thought that they
appear to promote. PowerPoint "helps you make a case, but it
also makes its own case: about how to organize information,
how much information to organize, how to look at the
world."[66] Perhaps the most notable detractor has been Edward
Tufte, the "analytical designer" whose books on envisioning
and displaying quantitative information have become bibles of
the field.[67] He has categorized presentation software as "slide-
ware" and in an essay entitled "PowerPoint Is Evil" asserts that,
"The standard PowerPoint presentation elevates format over
content, betraying an attitude of commercialism that turns
everything into a sales pitch." By "shaping presentations to a
standard format," according to some observers, and by "reject-
ing material that doesn't fit the format, the program controls
content, militating against the communication of complex ideas
in favor of bullet points projected on screen."[68] Furthermore, according to Tufte, "PowerPoint's pushy style seeks to set up a
speaker's dominance over the audience. The speaker, after all, is
making power points with bullets to followers."[69] The cover of
Tufte's booklet, *The Cognitive Style of PowerPoint*, is illustrated

with a photograph of a military parade in Budapest's Stalin Square, with the towering statue of Stalin saying, "Next slide, please."[70]

PowerPoint's bulleted slides are perhaps the most visual characterization of the eponymous presentations, but the manner in which speakers interact with the slides is surely its most detested. A familiar stance of a speaker using PowerPoint is to be watching the slides not as the audience does, that is, as projected on a large screen, but on the computer screen on the lectern. The speaker not only speaks to the computer screen but also points to the bullets and images on it. The audience is left out of the dialogue between creator and created. Many speakers also read verbatim their PowerPoint slides, thus committing the "sin of triple delivery, where precisely the same text is seen on the screen, spoken aloud, and printed on the handout in front of you (the 'leave-behind,' as it is known in some circles)."[71] Such slide-intensive lectures have been described as "Death by PowerPoint."[72] A *New Yorker* cartoon showed the devil (or at least a manager from hell) conducting an interview and saying to the prospective employee, "I need someone well versed in the art of torture—do you know PowerPoint?"[73]

Regardless of how PowerPoint desensitizes presenters or tortures audiences, its physical attributes are generally believed to make its use an unqualified improvement over 35 mm slide presentations. This is a shortsighted view, for not every aspect of a PowerPoint presentation can be said to be superior to the corresponding one employing a Kodak Carousel projector. The redesign of anything typically introduces new limitations at the same time that it boasts vast improvements over the superseded technology. According to one critic of technology, there is "the sense of wonder at seeing that a new form of technology actually works," but that sense "will soon give way to jadedness." He recalled that the first time he used a digital camera he "was amazed" that he could see the pictures immediately after he shot them. But, within a few days, he had identified its faults and failings and "had a list of ways the camera should be improved."[74] No wonder manufacturers release new models so frequently and so successfully.

Nevertheless, just as lantern slides and 35 mm slides coexisted for decades, so can we expect 35 mm slides and Power-Point to do the same, at least as long as the equipment to project the former remains operative. Those of my generation, who grew up professionally with 35 mm slides, straddle the fence between them and PowerPoint. Our library of slides is often too vast and familiar for us to want to contemplate digitizing it, and so we continue to lug loaded carousels through airports, hoping that there will be a working projector at our destination. Airport security asks if the curious circular shape showing up on the x-ray image of the contents of our roll-on is in fact a slide carousel, for the scanners "don't see many of them anymore." At the talk site, young audio-visual personnel, who have no trouble hooking up the most off-brand laptop to a digital projector, leave the operation of the slide projector to us.

Among the most regressive features of PowerPoint is its strict conformity to the rigid computer-screen format, which is roughly in the proportions of an $8\frac{1}{2} \times 11$ inch sheet of paper turned on its side. This television-landscape layout, which is contrary to the way most documents composed on the computer screen will be oriented when printed, was highlighted in a review of Microsoft's Windows 2003. Changes made to the e-mail program Outlook were praised for displaying a message so that it "looks less like a telegram and more like a handsome business document." The reviewer went on to note that, "This new layout exploits a nearly forgotten quirk of today's computer screens: they're wider than they are tall. Stacking window panes side-by-side makes infinitely more sense than the old vertical arrangement."[75]

This quirk of computer screens is not so readily taken advantage of in PowerPoint. Slides composed on it are mostly horizontal and must remain horizontal. Tall vertical images must be squeezed down into the horizontal frame, like basketball players in the doorway of a Frank Lloyd Wright house. This is a limitation that the old technology does not impose. With a 35 mm camera, slide images can be taken with the camera (and thus the film) in a horizontal or a vertical position, and they can be mounted in the landscape or the portrait mode. The square format of the slide mount enables both orientations to be projected to the same magnification, the opaque slide mount masking out the unilluminated part of the screen.

For use with 35 mm slides, older projection screens were designed accordingly, being of a generally square format when fully deployed. Engineers and scientists became accustomed to using the full area of such screens, mixing horizontal and vertical slides to advantage, employing, for example, the former for the full view of a suspension bridge and the latter for a view of its tower. The size of the projected slide images was optimized. (Architects and art historians who are accustomed to using two projectors to juxtapose slides for viewing two images simultaneously, often for comparative study, have been used to much larger screens, which frequently fill the entire front wall of a lecture room. I recently attended an art history presentation made with PowerPoint, and the pairs of images were stacked vertically and squeezed into a single computer-screen format.)

Business presentations, being mostly text and graphs, tend to consist of slides composed exclusively in the horizontal, landscape mode. This is, of course, the PowerPoint way, and it is naturally in conformity with the proportions of the computer screen. When projectors—not overhead projectors or slide projectors, but simply projectors, as they are usually called—became almost affordable in the late 1990s, Power-Point presentations squeezed out the old 35 mm slide show in more ways than one. Projecting the computer-screen image on a fully deployed square-format projection screen could never fill it up. There was annoying white-to-gray space left on top and bottom, like the annoying spaces above and below a letter-box movie shown on a typical television set. Deploying the projection screen only to the proportions of the computer screen eliminated the unsightly reminder of how tall a vertical image once could be projected. But if the screen did not have to be fully employed, then it did not have to be quite so long, and so office and school supply catalogs increasingly offered not the old square projection screens but newer ones in the golden section of the computer screen.

Projection screens have long been notorious for getting soiled, punctured, and torn. So in recent years, whenever a new screen has been ordered, it increasingly has had the new proportions. The latest one installed in the classroom in which I teach is perfectly matched to the image that emanates from the ceiling projector, so that an enlarged computer screen hovers above the class without any hint of a margin, disembodied from its surroundings like a phantasmagoria. Unfortunately, when I want to project 35 mm slides in this classroom, I have to reduce the image so that the vertical slides fit, which makes the horizontal slides considerably narrower than the screen. Whoever ordered the screen seems not to have acknowledged that anyone might want to use it for anything other than projecting an image of a computer screen, with or without a PowerPoint presentation.

New technology will naturally drive out the old, which it mimics in functionality and terminology, if the old is no longer readily available. But it does not follow that new technology is superior in every way to the old. It depends on how comprehensive and conscientious a proactive failure analysis was conducted prior to the redesign. Certainly 35 mm slides have the disadvantage of requiring film developing and processing, usually by someone other than the presenter, which takes planning ahead. In the case of PowerPoint, giving speakers "control over design decisions" eliminated this failing of the soon-to-be obsolete system, but the constraints of the computer screen impose a different failing on the new system, on top of all those concerns of Tufte and other critics over format constraining content. It is seldom that a new product is an improvement in every way for everybody. Its success is always fleeting. A challenger is always lurking in the shadows.

Like the magic lantern and digital projector, everything made and used by humans has been designed, in that it has been realized from an idea or its parts have been selected from the store of existing things, modified if necessary, and assembled into a new and purportedly improved thing. The components of any conscious design are deliberately arranged or assembled into a particular configuration to fulfill an intended purpose, but there are also accidental designs. Thus, an inventor can stumble across something novel and useful while trying to design something entirely different, the way Roy Plunkett discovered Teflon while looking for a new refrigerant.[76] There can also be a considerable amount of playfulness in the design process. Sometimes the purpose of a design is simply to amuse or entertain.

Components, like the designs assembled from them, need not be tangible things; they may have been chosen for their meaning or color or texture or shape or any other quality. Their final assembly may have been but one of many trial-and-error arrangements—most of which failed to satisfy their designer or those who used them—but the outcome is spoken of as *the* design. It has always been this way, and it always will be, for design is an activity independent of time and place. We can thus credibly imagine where things came from and know that they will evolve into ever-new designs.

Designs always beget designs. However, since design is a human activity, it is also an imperfect one. Everything designed has its limitations and its flaws. This fact of design is what leads to constant change in the things around us and our behavior involving them. Inventors, engineers, and other professional designers are constantly criticizing the world of things, which is what leads to new designs for new things. The successful new thing is one that does not fail in the way that what it is intended to supersede did. This is why failure is the key to design. Understanding how things fail—and might fail—provides insight into how to redesign them successfully. But today's successful design will be tomorrow's failure, for the expectations of technology are themselves constantly being revised.

Complex collections of things and our manipulation of and interaction with them are termed systems, which themselves are designed. We are embedded in systems of all kinds, some of which are obviously mechanical or electronic but many of which are also social and cultural. Thus, when we drive an automobile we become part of a human-machine system. At the same time, our behavior on the road is governed by rules and regulations, which are largely social and cultural. Driving for the first time in a country whose rules are different can be disorienting, but humans are supremely adaptable to novel technology. We learn quickly how to drive a (nearly) mirror-image car and how to do so safely on a crowded highway.

On college campuses with many international students, the first weeks of class often bring confusion on the footpaths, where those accustomed to walking on the left are encountered by those used to walking on the right. Such conflicts are readily and amicably resolved by flicking a smile and making slight corrections in our paths. That is what we are constantly doing with new designs embedded in an ever-evolving technology. But as much as things change, what drives that change remains essentially the same. It follows that any specific thing or system illustrates the nature of design.

2

SUCCESS AND FAILURE IN DESIGN

*Everything has two handles—by one of which
it ought to be carried and by the other not.*

—Epictetus[1]

Ur-implements were most likely those things found close at hand. Thus, the finger would have been the obvious choice to draw lines in the sand, and then to point out features of the plan. But the finger is blunt, and the hand to which it is attached obscuring. The humble twig or stick certainly became the natural extension of the finger, allowing the designer to stand away from a drawing and thus obscure it less. But even the act of picking up a stick can involve design. A good number of sticks lying about might be rejected as too short or too curvy, too flaccid or too brittle. They would not even have to be touched to be considered failures. A proper pointer might have had to be chosen from and broken off of a nearby bush or tree. Once chosen, it could have been held at either end—but the thicker generally feels better in the hand, leaving the more slender free to be more effective as a pointer. All these choices and decisions made in anticipation of the stick's use would have been ones of design.

An alternative, of course, was to shape a pointer out of a suitable piece of wood. In the nineteenth century, when the images projected by magic lanterns grew to great size, it took a larger and larger pointer to reach a detail that a lecturer wished to emphasize. This led to the making (and selling) of long, straight, and slender pointers out of appropriate pieces of timber. However, with considerable length came not inconsiderable mass, and a twelve-foot-long pointer could be rather unwieldy. Lecturers must have had a difficult time holding one at the very end, even for a short period of time, and the thicker end of the pointer might have had to be rested on the stage to be used without undue discomfort.[2]

The fabricated wooden pointer of more modest size was once standard equipment in the classroom. Though subtly tapered like a stick, it had the aesthetic and functional advantage of apt proportion. It was usually fitted with a rubber tip—to soften its tap and scratch on the blackboard—and a metal eye by which to hang it from a hook. The wooden pointer may be said to have been a "perfected" design in the twentieth century, but it had a propensity for being mislaid or being broken over a mischievous pupil and thus rendered less effective for its intended purpose. Even on college and university campuses, where corporal punishment is rare, wooden pointers had a habit of disappearing. Guest lecturers, wishing to stand away from the projected images they were describing, often were offered an ad-hoc pointer in the form of a disembodied automobile antenna, a yardstick or meterstick, a strip of slender wooden molding, or some other remnant from the lumberyard. These succeeded marginally as pointers but had severe aesthetic failings.

A Victorian lecturer using a large pointer during a presentation on the humors of parliament. (From *London News & Graphic*, 1891.)

More fastidious lecturers began to carry their own mechanical pointers, usually in the form of a purposely designed telescoping wand fitted with a pocket clip.[3] Collapsed, these implements looked like just another pen or pencil in the engineer's pocket arsenal. Unfortunately, in larger auditoriums, the limited length of these and most pointers required lecturers to stretch in the spotlight of their slides to tap things at the top of the illuminated image. (Some speakers held the pointer in midair between projector and screen to cast a shadow on the object of interest.)

The laser pointer, which came into prominence in the 1990s, was at first typically thick in the hand (to accommodate its electronics and batteries) but had a good range. Before long, such pointers came in more slender pocket models and in shorter, key-chain styles. The laser pointer has clear advantages over the simple stick, but it also has its own shortcomings. Batteries are required and they run down, the shaky hand of a nervous lecturer is amplified in the jerky motions of the glowing red dot, and the eyes of the audience are at risk. Moreover, the red dot can be difficult to see against certain backgrounds, thus rendering the high-tech pointer inferior even to its low-tech predecessors. To correct this problem, more electronically complicated and thus more expensive green-beam laser pointers, which reportedly can be as much as thirty times as intense as the red, came to be introduced.

Technological advances generally tend toward but never reach perfection; there is always some way in which they can be improved. Lecturers found themselves encumbered with a slide changer in one hand, a laser pointer in the other, and thus no hand left to turn the pages of their notes. On more than one occasion, I have seen lecturers press the slide advance button when they wanted to activate the laser beam, and vice versa. Slide changer controls came to be added to laser pointers, whose casing had naturally to be enlarged—into devices that were shaped not unlike the familiar domestic remote control, which most hands feel quite comfortable holding and manipulating. The laser pointer was thus at the same time evolving toward both intensification and complexification. But not every combination worked for every user.

Success and failure are the two sides of the coin of design. This is nothing new. Like the stick pointer, virtually all of the earliest things used in prehistoric times can be assumed to have been found in nature: caves in which to seek shelter, rocks with which to hunt (and fight), fallen branches to reach fruit high on trees, sticks to poke into beehives and insect holes, shells to scoop up water from a lake, fallen logs and stepping-stones to cross a stream. Though such found things may have needed no essential crafting, their mere selection for a purpose made them designed. Everything we have used since has also been designed, in the sense that it has been acquired, adapted, altered, arranged, or assembled deliberately to accomplish a specific objective. Designed things are the means by which we achieve desired ends. If the ends have not always justified the means, they have at least inspired them. But how do things evolve from sticks and stones to bricks and mortar? From shells to spoons? From logs to bridges? From caves to castles?

Whenever we use some *thing* to do *some*thing we expect it to do, we test it. Such testing is not necessarily conscious, but it is always effective and consequential. Indeed, with the testing of each individual example of a thing we also test the general hypothesis on which our expectation is based, whether consciously or not. If the thing passes the test, we declare it a success—at least until the next test. Successful tests are unremarkable. If the thing does not pass a test, we say that it (and the hypothesis) has failed. Failures are remarkable. The failures always teach us more than the successes about the design of things. And thus the failures often lead to redesigns—to new, improved things. Modern designers and manufacturers can do this on their own, or they can be encouraged to do it by consumers, who essentially are design critics who vote with their purchases.

I have a versatile piece of carry-on luggage that I have subjected to many tests. It is a roll-on that is sized to fit into the overhead bin of an airplane. The bag has a good many zippers, which give me access to a variety of outside pockets, one of which is expandable. When my carry-on was new, I often stuffed that pocket with books and folders, thinking that I would not have to carry a separate book bag. With the pocket so stuffed, the bag often failed to fit into the overhead. Such experiences taught me the limits to which I could expand the pocket, even though it itself had much more capacity. On the front of the bag there is a smaller pocket, which is not expandable but can be opened from either end. On one trip I put a book in that pocket and closed the top zipper but forgot to close the bottom one, which I had opened the night before to retrieve the same book. I only realized that my book had fallen out in the hotel lobby when a kind stranger ran after me with it. The bag functions as designed only when I use it as designed.

My roll-on has two conventional handles, one on the top and one on the side. I have found the top handle most convenient for carrying the bag up and down stairs, something I do often in taking commuter flights. The top handle is also good for carrying the bag down the aisle on larger planes. Early on, the way I lifted the bag into the overhead made the side handle inaccessible for retrieving it at the end of the flight. That had led to some torn fingernails. Now, I pay attention to how I stow the bag, making sure one of the handles is always facing out. Both handles are backed with Velcro, so that when they are not in use they hug the bag and do not get abused by the luggage of other passengers or by baggage handlers when I check my luggage.

Since it is a roll-on, my bag also has a collapsible handle by which I can pull it through airports and parking lots. When the bag is heavily packed, pulling it by this handle can get a bit uncomfortable. I have noticed that airline pilots and flight attendants, who almost invariably have this brand of luggage, tend to sling a second bag over the front rather than pile it on top, as I had tended to do. Now I follow the example of the professionals. The slinging method counterbalances somewhat the weight of the roll-on and thus lessens the downward force on the pulling hand, hence making it less likely for me to have to change hands frequently.

Once, when we were on a long trip, my wife bought a new roll-on bag. She did so in a store that carried a great variety, and so which one to choose among the many closely similar designs was not an easy decision. After narrowing down the choices to a few that had the right amount and arrangement of luggage space, she tested each of them by trying its handles and pulling the bag (empty) along the store's carpeted aisle. It was only after she had chosen and bought her bag and was rolling it back to the hotel that we realized that the wheels of her roll-on made a lot more noise than did mine, especially when pulled over a scored sidewalk. Furthermore, as I learned when I pulled her heavily loaded bag one day, its handle was shorter and thus closer to the ground than the handle on mine. This had not mattered so much when the bag was empty, but it did matter when the bag was full and heavy. The tests to which the bag had been subjected in the store did not represent the more trying conditions under which it would be used in an airport terminal. What had seemed to be a piece of luggage whose design differed from mine only in aesthetic ways proved to be functionally very different as well. Its performance was not up to what was expected, and we both have learned to be more critical shoppers the next time we buy a piece of luggage or anything else.

"Failure is an unacceptable difference between expected and observed performance," according to the comprehensive definition used by the Technical Council on Forensic Engineering of the American Society of Civil Engineers.[4] Good design is thus proactive failure analysis, something that both a designer and the chooser among designs ought to practice. Anticipating and identifying how a design can fail—or even just be perceived to fail—is the first step in making it a success. Still, whether we are designing or buying luggage, or building or occupying warehouses, we can overlook the details that make the difference between success and failure. Just as a roll-on need not fall apart completely to be a disappointment, so a building need not collapse catastrophically to be considered a failure. A warehouse with a door narrower than inventory it was built to store is a decided failure.

An "unacceptable difference between expected and observed performance" can result when a structure merely settles and cracks. In such a case, there can be considerable disagreement over how much should have been expected and how much should be considered unacceptable. Unfortunately, sometimes expectations are articulated only in retrospect—in a courtroom. Regardless, when a part of something engineered to be successful fails, the entire design has failed; and it is a signal for examination, for change, for redesign.

The new Walt Disney Concert Hall in Los Angeles is a striking piece of architectural and engineering design. Frank Gehry's imagination has given the hard stainless steel façade the appearance of soft and pliable flower petals. Shortly after completion of the building, however, an unexpected problem arose. The surface of one section—whose cladding had a bright finish instead of the brushed texture of the rest of the structure— reflected sunlight onto a condominium building across the street, blinding its tenants and raising the temperature in their living space by as much as 15 degrees Fahrenheit. Had the unacceptable result been anticipated, a nonreflecting finish might have been used. When I visited the concert hall, the offending surface was draped over with a netlike fabric to mitigate the unwanted effect while a permanent solution was being sought. That solution—or redesign—was, not surprisingly, to dull the bright finish, and it only remained to choose the process for doing so.[5] After almost a year of studying the problem, it was announced that the offending section would be sandblasted to "dull the finish so that it resembles the exterior of the rest of the building."[6]

Had the corner of the concert hall been oriented differently,

Some segments of the façade of the Walt Disney Concert Hall in Los
Angeles. (Photo by Catherine Petroski.)

the stainless steel cladding might not have reflected the sunlight into the condo or any other nearby building, and an issue
would not have arisen. The Disney Hall façade might have
been hailed as an unqualified success. Indeed, the critical acclaim for the structure might likely have reinforced Gehry's
conviction that juxtaposing bright and brushed stainless steel
produced the desired aesthetic effect. Without the corrective
required by the reflection problem, future Gehry buildings
might have been designed with increasingly daring variations
on the use of such cladding—perhaps drawing further acclaim.
As long as the façades did not concentrate excessive sunlight
onto broiling neighbors, the bright cladding might have been
used with increasing confidence, until what happened in Los
Angeles happened to some subsequent project. That incontrovertible failure would have revealed a fault that had only been
latent in all the earlier "successful" buildings.

The unintended consequence of reflected sunlight is only
one of myriad ways in which an architectural design can fail.
One winter, another new Gehry building—the Weatherhead
School of Management at Case Western Reserve University—
began to shed snow and ice off its sloping stainless steel roof,
potentially endangering pedestrians. The smoothly curving
surface provided no barrier to stop the accumulated frozen pre-
cipitation, once loosened, from arcing off the building like
skiers racing over a mogul. "The benefits of a striking building
outweigh a minor problem," one architect is reported to have
said. He also said, "If you're going to have that building shining
and beautiful and doing all the wonderful things it does, this is
a small price to pay." Not everyone agreed. Among the fixes
contemplated in Cleveland was the installation of heating

cables to prevent the snow and ice from building up in the first place.[7]

Getting things just right can be tricky in any context. When a new commuter car—designated the M7—was being developed for the Long Island Rail Road, passenger complaints about earlier car designs were taken into consideration. After many years of service, the neoprene seats found on the earlier M1 and M3 model cars had "lost much of their bounce" and developed "sinkholes," which were blamed for backaches. So the seats on the M7 were "injected with silicone, given lumbar support and a headrest" to improve them ergonomically. Riders had complained also about the absence of an armrest for the window seat, and so in the redesign "one was molded into the wall of the car." Focus group members who otherwise liked the seats on another model car, the double-decker C3, reported of it that "they found the armrests just a tad short." So the armrests on the M7 were made four-tenths of an inch longer than those of the C3. Yet this seemingly insignificant but well-intended modification made by the designers "snagged their dream of design perfection on the railroad cars of the future." Within two years of the M7's introduction, "73 claims for torn clothing had been filed," blaming not only the extra length but also the "rubber eraser"-like material of the new armrests for catching trousers and not letting go when the unlucky commuter got up from his or her seat.[8]

Blaming an unfortunate occurrence on bad design may make for a convincing damage claim—or even a successful lawsuit—but the connection between intention and result, between cause and effect, is not always what it seems. Over three thousand intersections in New York City have signs instructing pedestrians, "To Cross Street / Push Button / Wait for Walk Signal."[9] A good deal of time often elapses between pushing the button and getting the go-ahead, but conscientious citizens obediently wait. They presume, one presumes, that a delay is part of the system's design. It may be a "bad design," but the light does change—eventually.

New York intersections began to be fitted with these "semi-actuated signals" around 1964. They were the "brainstorm of the legendary traffic commissioner, Henry Barnes, the inventor of the 'Barnes Dance,' the traffic system that stops all vehicles in the intersection and allows pedestrians to cross in every direction at the same time." Walk buttons were installed mostly where a minor street intersected a major one, along which traffic would be stopped only if a pavement sensor detected a vehicle waiting to enter from the minor street or if someone pushed the button, causing the light to change ninety seconds hence. With increased traffic (by 1975, about 750,000 vehicles were entering Manhattan daily), the signals were being tripped frequently by minor-street traffic. The walk buttons hardly seemed necessary, and pushing them interfered with the coordination of newly installed computer-controlled traffic lights along many thoroughfares. Consequently, most of the devices were deactivated by the late 1980s, but the buttons themselves and the signs bearing the instructions for their use remained in place. Evidently there was never any official announcement about the status of the "mechanical placebos."[10]

Though most New York City traffic-crossing buttons are thus "off" when they appear to be "on," many familiar consumer electronics devices in fact remain "on" when we turn them "off." This passive design feature is necessary so that our stereos, televisions, and other home appliances can respond to the touch of a keypad or the signal from a remote control. According to one study, a sizable percentage of microwave ovens "consume more energy in standby (running their clocks and keeping their touch panels active) than in cooking" over the course of a year. This phenomenon has come to be termed "standby power waste" and amounts to as much as 10 percent of electricity used in U.S. homes.[11]

Similarly bizarre design quirks arise when computers are really "on." In 1998 IBM incorporated a very useful feature into its ThinkPad: The keyboard can be illuminated, which is obviously handy when typing in poor light, such as in the darkened cabin of an airplane. However, because there was no place to put an on-off switch or button or key dedicated to the lighted-keyboard feature, it is activated by pressing the unintuitive combination Fn+PgUp. Even though there is a lamp icon on the PgUp key, evidently few users knew its significance or that the keyboard could be lighted up, and so they continued to type in the dark. According to an IBM director of design, "Not everything can have a button. If we didn't limit them, we'd be looking at products with as many buttons as an accordion."[12] Judgment calls pervade design.

Like these, most failures and our responses to them are not matters of life and death. In fact, most failures are fairly innocuous shortcomings that are generally not viewed with any great alarm by either maker or user. Computers and their software are notorious for inexplicably locking up while performing a task. Everything might seem to be working fine, when something goes amiss and everything stops. There is no response to hitting any key, and even holding down the

Ctrl+Alt+Delete keys does not appear to do anything. In such cases, the only recourse seems to be to pull the plug or remove the battery, hoping that all will not be lost when power is restored. The failure is often blamed on some unnamed hardware or software "bug," a term that suggests something at the same time unwanted but ubiquitous, generally innocuous but potentially deadly. A bug can make the computer "crash," but there is no visible physical damage. When this happens, we have become accustomed to just starting the machine up again, hoping that whatever was the problem has gone away. The phenomenon is so common that it has become encultured in the form of jokes that everyone gets. According to one, a physicist, an engineer, and a computer scientist are riding along in a car, when suddenly it begins to smoke and its engine stops. The physicist suggests the problem to be a theoretical one involving torque; the engineer suggests it to be a mechanical one that can be fixed with an adjustment; and the computer scientist suggests that they all get out of the car, wait a minute, and then get back in and try to restart it.[13] In fact and in all seriousness, even physicists and engineers resort to this solution. "Many a user reboots his or her personal computer routinely," and thus, according to one approach, "engineers should design systems so that they reboot gracefully."[14]

Rebooting the entire system is considered to be "the most common way to fix Web site faults,"[15] regardless of their source. One "breakdown of typical failure causes for three Internet sites" attributes 15 percent to hardware errors, 34 percent to software problems, and the remaining 51 percent to operator error. How to deal with the main cause is not obvious:

When designers of other engineering systems have discovered a propensity of operator error, they often have attempted to remove the need for human input. Removing human operators can lead to a well-established pitfall known as the Automation Irony. Because designers can typically reduce but not eliminate the need for human intervention, such efforts frequently make things worse. That's because engineers generally automate the tasks that are easy, leaving the hard jobs for people. These measures mean that administrators must carry out difficult tasks intermittently on unfamiliar systems—a sure recipe for failure.[16]

The phenomenon is similar to that remembered by an engineer who began his career before the pervasive use of computers. He was advised by his boss that "it might be possible to make a thing 'foolproof,' but probably not 'damnfool proof.'" Furthermore, the boss hoped that the younger engineers would not try to "improve things worse" with endless design changes.[17] Today, we might say endless upgrades.

Computers are arguably among the most complicated and mysterious things that consumers buy and operate. Uncountable, invisible electrons conduct instructions instantaneously from keyboard to screen, on which we can manipulate words, numbers, and images as if by magic. Compact DVDs smaller than an old 45-RPM record that would hold just two short singles now hold all the sounds and sights of a Broadway musical—and give them up to the screen at the click of a mouse. To do all that it does, Windows 2000 was designed with twenty million lines of source code, and Windows XP with forty million.[18] For all their idiosyncrasies and faults—and features known and unknown—we trust computers, as we do our confidants, with our most private thoughts and count on them for our most critical financial calculations. We know computers have bugs and do not work perfectly, yet when they do work we assume that our words are securely saved and our numbers are reliably added. So basic an assumption is not necessarily justified.

Hardware bugs tend to be the least common because "designers of microprocessor chips have regularly added circuits to simplify the testing of chips, even though these additions increase chip size and remain unused after the microprocessors leave the factory." Nevertheless, this approach has been considered cost effective "because the test circuits allow designers to inject 'failures' artificially to verify that the chip detects and recovers from them correctly." It has been proposed that computer system designers devise analogous software "means to inject errors to test systems' ability to return to service," thus providing a way to compare the dependability of different systems.[19]

What has been called "the most widely publicized hardware bug in computer history" was discovered in 1994 by a professor of mathematics who was using the then-new Pentium processor in his work on prime numbers. Under certain conditions, he found, multiplying two numbers produced incorrect results. At first he did not suspect the new computer chip and looked elsewhere—including to his own work—for the error. Puzzled by the mystery, other mathematicians, computer scientists, and engineers also spent inordinate amounts of time trying to figure out what was going wrong. Intel, the manufacturer of the chip, had itself discovered the bug earlier but had not volunteered information on what it called the "subtle flaw." Rather than issuing a recall, Intel continued to sell the imperfect chip. When it finally did admit to the problem, which in the meantime it had corrected, the company required "owners of Pentium systems to justify their needs for a replacement chip," by demonstrating that the kinds of calculations they did demanded sufficient precision to make a difference. This position angered many users and consumer advocates, and eventually Intel succumbed to the pressure and agreed to provide replacement chips just for the asking.[20]

The Pentium bug issue became highly publicized in large part because of the persistence of sophisticated computer users and because of the way Intel mishandled the public relations of the matter. At the same time that it was in the news, Microsoft was getting ready to distribute "final beta" copies of the operating system Windows 95. One critic anticipated that, "Even with years of testing, and potentially more than a million test users, there will be bugs in Windows 95. It's a given."[21] That may be true, but Bill Gates has been quoted as saying, "There are no significant bugs in our released software that any significant number of users want fixed."[22] This may also be a given, since "consumers have become so accustomed to flawed software" that patches and upgrades can be issued without overly angering users.[23]

With the development of the Internet and the World Wide Web, the capabilities of this thing known as my laptop, with all its flaws, have multiplied virtually without bounds. When I once asked Google to find something for me, it reported that it had "searched 4,285,199,774 web pages" in 0.24 seconds to return "about 203,000" results in response to my query relating to the introduction of New Coke. (A year later over eight billion pages would be searched.) Perhaps it is no wonder that every now and then my computer freezes up, just as an individual might under the thought of completing such a task so quickly. Ignorable failures, like individual computer crashes, are expected and commonplace and, so, unremarkable. As we do when the keyboard freezes up, we accept this (for we have no other choice) and start over again. We do this with varying degrees of anger and varying displays of disgust, depending upon the task at hand and the deadline under which we are working, but usually we just reboot the computer and continue with our work. Generally, we do not even bother the manufacturer, and so it may not even be aware of how widespread the problem might be in its product—though that may seem unlikely. But colossal failures, like the crash of an entire large computer network, are relatively rare, and that is why they are headline news.

There is a group of people who do not readily dismiss, let alone accept, even the most trivial of failures. Indeed, they see failures where most of us see only successes. These are the inventors, the engineers, the designers of the world, who are forever trying to improve it through the things in it. To these intrepid pioneers of purpose, a failure of any kind is not so much a disappointment as an opportunity. They tweak the things we know to turn them into things we did not even know that we needed. In conjunction with sales and marketing people— themselves designers of a sort—they change the world a thing at a time. And the single most common characteristic of such creative people is the way in which they view failure. They recognize that a failure not only provides them the opportunity to carry out the process of design and development anew but also enables them to conceive of something new and improved to obviate the triggering failure. According to Ralph Baer, who has been called the "father of the home video game," inventors share the characteristic of "looking at the world as if everything in it needs fixing."[24]

A young Thomas Fogarty found something that needed fixing while working as a surgical scrub technician in the 1950s and witnessing what was then the standard operation for removing a blood clot from a blocked artery in a leg. The procedure involved making a long incision, in some cases extending from the abdomen to the knee, to expose the clotted blood vessel. The time-consuming and complicated operation had a large failure rate and could result in amputation or death. The desire to reduce the incidence of failure in this procedure led Fogarty to devise the world's first balloon catheter, which he patented in 1969.[25] He went on to invent and patent many other medical devices and in the process had a lot of experience with false starts. Like many an inventor, he took such setbacks in stride, recognizing that "failure is the preamble to success."[26]

Failure is not only the preamble but also the way to success in design. Failure of an existing thing or technology provides not only the initial motivation for seeking an improvement in some thing or some process but also the means of incrementally developing ideas and prototypes into patentable inventions. Fogarty's early prototype for the balloon catheter was made by employing fly-tying techniques from fishing to attach to a plastic catheter the pinkie tip snipped from a latex glove.[27] Jury-rigged prototypes are important for "debugging" designs.

Prototyping can be viewed as a "sort of three-dimensional sketch-pad," with the prototype enabling potential backers and users to see the invention as a tangible thing. Dennis Boyle, a studio leader at the design firm IDEO, further sees the construction of "rapid and rough prototypes" as a means of identifying problems early in the design process, when they are less costly to correct. According to Boyle, if a "project is not generating masses of prototypes, including many that clearly won't fly, something is seriously wrong." The creed at IDEO is thus "Fail early, fail often."[28]

The sentiment is not new, nor is it unique to design. According to Samuel Smiles, the Victorian biographer of engineers who wrote about their overcoming great personal and technical challenges, "We learn wisdom from failure much more than from success. We often discover what will do by finding out what will not do; and probably he who never made a mistake never made a discovery."[29] The American poet James Russell Lowell conveyed a cognate idea by means of a simile: "Mishaps are like knives, that either serve us or cut us, as we grasp them by the blade or the handle."[30] What was true for engineers and poets—and everyone—in the nineteenth century remains true for everyone today. As the design critic Ralph Caplan has written, "The more things change, the more we stay the same."[31]

How individuals react to failure separates leaders from followers, true designers from mere users of things. Professor Jack Matson of Pennsylvania State University believes so strongly in the role of failure in design that he expects students in his Innovative Engineering Design course to fail in order to pass. The course, nicknamed "Failure 101," requires students "to build and attempt to sell outlandish and frequently useless products," like a hand-held barbecue pit. The most successful students in the course are those who take the most risks and so fail the most. Matson hopes to get them to the point "where students learn to disassociate failures resulting from their attempts to succeed from being failures themselves." He believes that, "Innovation requires that you go beyond the known into the unknown, where there might be trap doors and blind alleys. You've got to map the unknown. You map it by making mistakes." It is not unlike being blindfolded in a labyrinth. Smacking into walls may signal a misstep, but the sum of those missteps defines an outline of the maze. The quicker more mistakes are made, the quicker the maze is mapped. Matson is an advocate of "fast failure."[32]

Whether fast or slow, failure and its avoidance have always been central to the development of designs and their far-reaching influence. Though often considered apocryphal, the familiar story of the standard railroad gauge of 4 feet 8½ inches serves as an example. This odd distance between rails is believed to be rooted in ancient times, when all Roman war chariots came to have that same wheel spacing, which is said to have been established to be no wider than the rear ends of the two horses that pulled a chariot. This width, which prevailed throughout the Roman Empire, ensured that the horses would not pull a too-wide wagon through an opening only wide enough for them. As the standardized chariots ranged throughout the empire, they wore deeper and deeper ruts in the Roman roads, including those in England. So, the development of English wagons incorporated the same wheel width, lest their wheels not ride in the ruts, the path of least resistance and least damage—and of least failure. Early tramway and railroad cars were built employing the same wheel width because they were made by wrights using the same patterns and tools as they had used for horse-drawn wagons. The engineer Robert Stephenson, who by 1850 would be "involved in the provision of a third of the nation's railway network,"[33] adopted what came to be known as the "standard" gauge.

The legendary British engineer Isambard Kingdom Brunel was certainly not alone in recognizing the arbitrariness of laying out a nineteenth-century railroad according to what has been purported to be an ancient Roman measure, and rather than accept the "standard" he designed his Great Western Railway with a gauge of seven feet, which came to be known as the "broad" gauge, in contrast to the narrower one. After all, the railroad cars would never have to run on old Roman or even English wagon roads. By adopting a broader gauge, Brunel believed he could design railroad carriages that would be more commodious, would provide a smoother ride for passengers, and could be pulled faster. In 1845 Britain had 274 miles of broad gauge track and 1,901 miles of narrower track, but its railroads could not easily continue to develop with incompatible gauges. A royal commission was appointed to study the problem, and it recommended that the Stephenson narrower gauge become the standard, arguing that it was easier and more sensible to convert everything to it. Since the early railroads in America followed the English model, the odd gauge became the U.S. standard also.[34] It was not the world standard, however. Through the twentieth century, the "Irish Standard" gauge—used also in southern Brazil and southeastern Australia—was 5 feet 3 inches. This clearly impeded the import and export of rolling stock and prevented the development of a railway ferry system across the Irish Sea. In 2000, by one count, there were eight "principal railway track gauges" throughout the world.[35] In design, which is more art than science, the logical thing to do is not always the practical or politic thing to do—and vice versa.

The more track was laid to the standard gauge, the more it became virtually impossible to challenge it as the default design. The implications of this design imperative have been far reaching, affecting even space travel in the late twentieth century. Indeed, the design of the solid rocket boosters essential to launching the space shuttle was strongly influenced by the standard railroad gauge. Since the boosters were manufactured by the Morton-Thiokol Corporation at its plant in Utah, they had to be designed in such a way that they could be shipped by rail across the Rocky Mountains and on to the Kennedy Space Center in Florida. The railroads passed through the mountains via tunnels, whose width was naturally tied to the track gauge. Thus, the booster rockets had to be shipped in sections no larger than could fit through the tunnels. Once at the launch site, they could be assembled into full-size boosters by incorporating the O-rings that proved to be the design detail that initiated the *Challenger* failure.[36]

Even though an initial design choice may have been rooted in some arbitrary decision, it is often the case that so many things have followed from that choice that to change any one of them would be to cause a concatenation of failures. The first use of 35 mm film was in making motion pictures, an enterprise that Thomas Edison carried out at his West Orange "invention factory." According to Kodak company lore, George Eastman asked Edison how wide the film had to be for his new movie camera. Edison is said to have held up his thumb and forefinger separated by approximately 35 mm and said, "About this wide." The story is most likely apocryphal, since the film a camera used would obviously have had to be more precisely specified. George M. C. Fisher, erstwhile chairman of the Eastman Kodak Company, has offered a different explanation: "It's more likely that a roll of 70 mm film, which was common in the 1890s, was slit in half to make 35 mm film. If this is the true source of the standard, it's just as arbitrary as the legend."[37] Neither 35 mm film nor the cameras that use it could be altered in disregard of the other, but sometimes one component of a design pair can be changed without altering the mating part. Nonetheless, while the result might work functionally, it can still fail aesthetically.

Aleve, an over-the-counter form of the prescription pain reliever Naproxin, was introduced in 1994.[38] Ten years later, most users of the little blue pills had become accustomed to finding them packaged in a "child-resistant" plastic bottle whose top could not be unscrewed unless the fingers of one hand squeezed the little tabs on either side of the bottle proper at the same time that the other hand turned the cap counterclockwise. The patented bottle itself is of a flattened shape, which helps keep it from twisting in the hand holding it.[39] However, the overall appearance of the bottle, which was protected by a separate design patent, could be said to have been compromised by the inclusion of the tabs, which introduce awkward and unexpected breaks in the otherwise smooth white surface.[40] Nevertheless, they clearly did serve an important function and became associated with the brand's packaging.

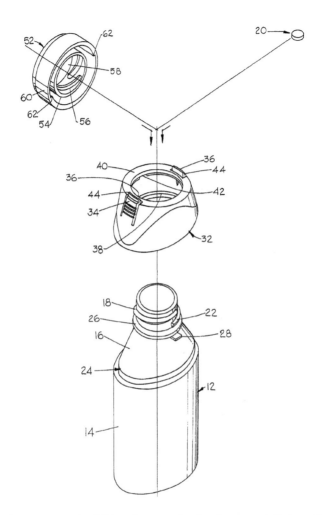

One improvement in child-resistant packaging for potentially harmful drugs comprised a two-part bottle design. (From U.S. Patent No. 4,948,002.)

Although the Aleve bottle was designed to be "readily open-able by adults, particularly adults having impaired manual dex-terity of their hands and/or fingers,"[41] as would the many arthritics taking the pills, it evidently did not prove so easy for all of them to operate in practice. Rather than lose an impor-tant segment of the market because in their estimation the packaging was a failure, the manufacturer designed a new "easy open arthritis cap" that received an "ease-of-use commenda-tion for packaging" from the Arthritis Foundation.[42] Unlike the old cap, which had a smooth edge that matched the texture of the body of the bottle, the new cap had prominently raised ridges that enabled a better grip to be applied. An even more significant change was hidden on the inside of the cap: It did not have the cam- or toothlike protrusions that engaged the tab extensions and locked the old cap in place. As a result, the new cap was unlockable and could be easily unscrewed from the bottle, which now bore the caveat, "package not child-resistant." With the new cap, the tabs on the side of the bottle were unnecessary, of course, but they remained there nonethe-less, along with their now functionless extensions, vestiges of the original cap and bottle system. Another vestige was the warning on the label, which read in part, "Do not use if foil seal imprinted with 'Safety *SQUEASE*®' on bottle opening is miss-ing or broken." It was, of course, the "Safety *SQUEASE*" fea-ture itself that was missing.[43]

It was no doubt less costly at the time for the manufacturer of Aleve to use a single bottle design (and foil seal) regardless of which cap was being affixed on the assembly line, but this was done at the expense of aesthetics, consistency, and design integrity. The new Aleve bottle-and-cap combination was an

Child-resistant (*left*) and more easily opened "arthritis" caps (*right*) have been used on otherwise identical bottles of Aleve pain reliever. (Photo by Catherine Petroski.)

affront to good product design (and common sense) and an example of how designs and redesigns, through piecemeal or stopgap responses to problems, can degenerate from success to failure. Clearly the child-resistant Aleve bottle (with pushtabs and label and seal) was designed to go with the child-resistant cap; they were a "package," according to the patent that protects them.[44] Strictly speaking, a different bottle (and label and seal) should have accompanied the "easy open arthritis cap," to avoid the conflict of purpose and signal.

As I discovered when I bought a package of Aleve, the

conflict did not stop with the bottle, which was a good inch shorter than the box it came in. This resulted in a rather noisy package. When shaken longways, the bottle shuttled back and forth, hitting the ends with a thud that combined with a rattle from the loose pills (Aleve bottles contain no cotton, presumably another attempt to aid arthritics).[45] The 100-caplet bottle was packaged in the same-size box that can hold a larger (bonus) 130-caplet bottle, which I had also once bought. (In fact, several different quantities of Aleve come in the same-size box, which makes shopping for the pills confusing.) The mismatch of bottle and box size, like that of ribbed top and notched bottle, also signals an inappropriateness of design.

Further inspection of the Aleve packaging revealed signs that neither did its creators pay close attention to details of graphic design, or at least they did not demonstrate a strong enough critical sense about consistency. The top of the child-resistant cap was imprinted in its center with a large pair of irregularly shaped arrows chasing each other in a counterclockwise direction and encircled with the words "SAFETY SQUEASE." Another, smaller pair of arrowheads was located on the edge of the top, and they pointed outward toward the safety tabs on the bottle proper. Between these arrowheads were the instructions, "SQUEEZE TABS AND TURN CAP." One had to infer that the direction in which the chasing arrows pointed was the direction in which one was to turn the cap to open it. One should not have inferred, however, that the outward pointing arrowheads indicated that the tabs were to be pulled (rather than squeezed) in that direction. Overall, the unbalanced and inconsistent (and opposite pointing) imprinted directions, if they may be called that, should have been considered a graphic and instructional failing calling for a re-design. The graphic design of the directions imprinted on the top of the arthritis cap was much more direct. The two arrows chasing each other were of a more graceful, three-dimensional design, and the words "easy open cap" encircled them in a more graceful way. Understandably, there was no mention of the vestigial "squease" tabs, a distracting detail with no aesthetic or functional purpose. This, too, was a failing crying out for a redesign.

It was the failure of the child-proof Aleve bottle to be opened easily by arthritics that evidently lead to its redesign, however clumsy it was. The new cap, in turn, led to an additional caveat being printed on the already graphically cluttered box: "This package for households without young children."[46] That statement was absent from the latest bottle of Aleve that I bought at a warehouse club.[47] This package contained 250 caplets in a bottle that had still another cap design, which did bring some consistency to the package. The straight-sided and knurled bottle cap was of the familiar kind that you "push down & turn to open." The bottle itself had the now-classic Aleve patented shape, but without the squeeze tabs on its sides. Without them, the bottle could be and was made in a single piece, eliminating the horizontal assembly seam that might have been considered another blemish by design purists.[48] Finally, the seal was an updated one that read simply, "Sealed for your protection," thus eliminating the last trace of the original "Safety *SQUEASE*" design. The new bottle-and-cap combination has a much cleaner look, and one that I hoped would remain stable for a while. While the arthritis-cap design had evolved in a positive direction functionally, it had gone in a negative direction aesthetically. In this regard, it had diminished the overall design landscape of countertops and medicine cabinets, which generally need all the help they can get.

Though not always totally successfully, inventors, product developers, and manufacturers are constantly addressing failure, whether it be formal, functional, or financial. Hence we find a preponderance of items on supermarket shelves bearing the designations "improved recipe," "less fat," "lower calories," "better tasting," and the like. The criteria may or may not be different for products on the pharmacy shelf, but they are no less comparative: "improved formula," "faster acting," "longer lasting relief," "better tasting," etc. The field of consumer electronics has been especially competitive and comparative, with advertising touting ever "faster," "smaller," "thinner," and "lighter" portable devices. All competing products—all competing and evolving designs—are necessarily compared with each other and their predecessors, and all such comparative claims are implicitly calling attention to the competition's—or a predecessor product's—failings.

The familiar and trademarked Band-Aid dates from the early 1920s, but the basic idea had a long prior history. In the midnineteenth century, bandages spread with a poultice or liniment to be applied to a wound were known as mustard plasters or, simply, plasters. When the plaster came with an adhesive strip to hold it in place, it was known as a sticking or adhesive plaster. However, the technology left much to be desired. In an 1845 patent for an "improvement in adhesive plasters," William Shecut and Horace Day touted their design's porosity, which allowed "perspiration and other morbid matter" to escape from the covered wound. They claimed their improvement to be "infinitely superior" to existing adhesive medicated plasters, which they indicted for "being stiff and hard and apt to crack, and when worn any length of time corrugate in wrinkles or folds which cause much uneasiness." Furthermore, midnineteenth-century adhesive plasters were "extremely difficult to remove." In contrast, Shecut and Day's plasters were "always soft, adhesive, and porous."[49] A century later, users of Band-Aids (some users of which even in America called them "plasters") might have wondered if the technology had in fact advanced or regressed.

The Johnson & Johnson Company was founded in the mid-1880s to make newly developed antiseptic surgical dressings. As the company's scientific director, Fred B. Kilmer (the father of the poet who wrote "Trees") was responsible for dealing with product complaints. Kilmer's response to complaints about "skin irritations caused by the Company's plasters" was to package them with talcum powder. The talc soon came to be sold by itself as baby powder.[50] The Band-Aid was invented under Kilmer's watch, but not at the company. In 1920 Earle Dickson, a cotton buyer for Johnson & Johnson, was living in New Brunswick, New Jersey, with his new wife, Josephine. She was prone to cutting and burning herself in the kitchen and when home alone found it difficult to bandage her own wounds with separate pieces of gauze and adhesive tape. Dickson got the idea to make some "ready-made bandages by placing squares of cotton gauze at intervals along an adhesive strip and covering them with crinoline."[51] The first Band-Aid Brand Adhesive Bandages, which went on sale in 1921, were three inches wide and eighteen inches long, certainly intended to be cut to the size and shape needed.[52] (I bought a similar, but much longer, "dressing strip" in Canada a few years ago. It was sold under the British-trademarked brand name Elastoplast, and I found it quite convenient to use, as long as a pair of scissors was handy.)

Some of the Band-Aids in my medicine cabinet today are of the "flexible fabric" kind, and the box they come in touts their improvements over other brands and styles: "Extra Flexibility . . . Non-stick Pad . . . Greater Durability." This Band-Aid "moves with you to fit better . . . won't stick to wound for gentle removal . . . stay[s] in place longer." This is not hype, for I have found these Band-Aids to be everything Johnson & Johnson claims—and more. Once I applied one of them to a bathroom rug to mark a stain for treatment when the rug was washed. The Band-Aid stuck so tenaciously that it could be removed only by tearing away small fragments of it at a time, but these pulled tufts of rug with them. Rather than risk making a hole in it, we put the rug in the washing machine with the hope that the Band-Aid would float off in the water. That did not work. It was only after going through the dryer that the Band-Aid sloughed reluctantly off the rug. A thing that works well as designed can be a nuisance when put to a use for which it was not intended.

Virtually all manufacturers make claims about the performance of their products. When a new line of vacuum cleaners was being developed for Montgomery Ward, "designers began with a study to secure data on the features of all competitive products." As reported in a 1940 issue of the trade magazine *Electrical Manufacturing*, the results of the study led to a series of design goals, which included "increased cleaning efficiency, better appearance, low cost, reduced operating noise, convenience in operation, lighter weight and maximum accessibility for servicing."[53] The list of implicit and explicit comparatives could easily have been rephrased as a list of failures of the competition: inefficient cleaning, poor appearance, high cost, loud operating noise, inconvenient operation, heavy, and poor accessibility for servicing. Though it has been said that in nature "there are no failures, only feedback," in design and manufacturing there are both.[54]

Guaranties and warranties are acknowledgments of the central role that failure plays in the minds of manufacturers and consumers alike. What is being guaranteed against is, of course, the possibility of failure. The Zippo cigar and cigarette lighter was long advertised to light "first time, every time," an implied guarantee against its failing to produce a flame, even in the wind.[55] A 100,000-mile warranty on a new automobile is the manufacturer's way of implying that the warranted parts, at least, will not wear out before that amount of use. By implied extension, the car itself will be running at that time. Otherwise, it would not be competitive.

The veteran venture capitalist Michael Brown, in an imaginative essay set in the year 2023 in which his children and grandchildren reminisce about him, is quoted as explaining how the ultimate success of fuel cells was dependent on their achieving good reliability. According to Brown, "When fuel-cell manufacturers could count on only one failure in about 5,000 hours, and when they could guess what would fail, they were able to provide a warranty for that period. They would also pay for any extra work needed on other mechanical components if you had to visit the dealer more than once every 50,000 kilometers. Even the best old-style cars can't do that."[56]

The reliability of conventionally powered "old-style" automobiles has already become remarkably high, as is evident every time we drive hundreds of miles along an interstate highway without seeing a single disabled vehicle by the side of the road. When we do see a car with its hood raised it is typically an older model, reminding us of how common it was just decades ago for a fan belt to break, a hose to spring a leak, or an engine to overheat. Reliability (the opposite of failure rate), as expressed through warranties, has become an essential marketing tool for winning customers over to automobile brands.

But freedom from overt failure is not sufficient for success, as the story of the Oldsmobile shows. In 2000 General Motors announced that the make that had been produced since 1903 (with some annual outputs exceeding a million individual cars) would soon no longer be made. It was not that the Olds line of automobiles was not reliable, but that the company, which once sported the powerful Rocket V-8 engine and introduced early versions of automatic transmission and front-wheel drive, was simply not able to find its niche in the wake of the 1970s fuel shortages. Smaller models of the make whose name had for so long been associated with large cars could not hold the market for Olds. Its demise, which finally came in 2004, was blamed in large part on the General Motors "practice of applying different trim to lookalike and runalike GM cars," a practice known as "badge engineering." Without a clearly distinctive identity that enabled Oldsmobiles to be advertised as larger, faster, or in some other way distinctly better than the competition, they were left behind.[57] Oldsmobiles had failed to hold their own in a market where image is, if not everything, something close to it.

Coca-Cola once learned a similar lesson. After World War II, the company lost market share for years, falling from 60 percent just after the war to 24 percent in 1983. Much of the loss was the gain of arch competitor Pepsi-Cola, whose sales continued to rise while Coke's flattened out. Evidently as part of a strategy to alter the trend, in 1985 the company introduced New Coke with the superlative comparative slogan, "The Best Just Got Better." According to one report, the "New Coke formula consisted of the Diet Coke formula with real sugar replacing saccharine";[58] according to another, it was simply "a sweeter, more Pepsi-like version of the original."[59] Whatever it tasted like, cola drinkers did not like New Coke, and they especially did not like the fact that within a week of its introduction the production of original-formula Coca-Cola had ceased. Consumers were angry that they did not even have the option of going back to their old favorite. Within three months of its bold action, Coca-Cola brought back the discontinued formula as Classic Coke, admitting a mistake: "all the time and money and skill poured into consumer research on the new Coca-Cola could not measure or reveal the deep and abiding emotional attachment to original Coca-Cola felt by so many people."[60] Conspiracy theories abounded, with some people believing that "the deliberate purpose of introducing New Coke" was to bring increased attention to and renewed demand for the classic formula. "Others believe New Coke was a distraction to keep Americans from noticing the switch between cane sugar and high fructose corn syrup" in the original formula.[61] The general consensus appears to be that the company simply had failed to appreciate that what it saw as "an unacceptable difference between expected and observed performance" in market share, the consumer would see in taste and tradition. Failure takes many forms and plays many roles.

3

INTANGIBLE THINGS

Wit's an unruly engine, wildly striking
Sometimes a friend, sometimes the engineer.

—George Herbert[1]

It is not just the tangible things overtly associated with engineering and technology that are invented and designed—and that succeed and fail. Lectures and books have to be designed, anticipating as much as possible the expectation of the audience. Critics are nothing if not failure analysts. Coming up with a successor to a success is no mean feat. Many a critically acclaimed first novelist has produced a disappointing follow-up book, or one that is but the first just slightly redesigned and repackaged.

Systems of all kinds are also designed, in a big way. As I have written elsewhere, our system of shelving books is not obvious: In early chained libraries, the shelves were not flat and horizontal, the books were not vertical, and their spines were not visible. The modern custom of shelving books vertically on level, horizontal shelves with the spines facing outward had to be invented. Its components had to be designed, in stages, as did the tangible infrastructure to support its use. The complete modern system of shelving books was not widely adopted until well into the seventeenth century, and even then there remained libraries that did not conform to the new way.[2]

Now, of course, how we shelve books is taken for granted, as if it were always done the way it is today. Even more recent examples of system design, like the electronic marvels known as personal computers and related devices, are treated as if they have always been around. Still, they are constantly undergoing redesign. In 2004 Apple introduced an updated version of its iMac in the form of "a minimalist all-in-one system that hides the computer's internal components inside a flat-panel L.C.D. display." At the time, the company's head of design, Jonathan Ive, remarked that the goal of redesigning the iMac had been to "make it appear extremely simple," which resulted in "a display screen that balances on a thin metal stand and that can function with only a single power cord." According to Ive, the solution was "so essential and so inevitable that it seems like it wasn't designed" at all.[3]

But everything, including nations, governments, and constitutions, has to be invented and designed. The musical question "Who invented Brazil?" opens the 1933 *marcha* "História do Brasil," by Lamartine Babo. According to one accounting, "Brazilianness was commonly understood to mean that collection of qualities which defined the nation, which distinguished Brazilians from citizens of Argentina, Portugal, and the United States—to name three populations whom Brazilians felt it was important to define themselves *against*."[4] The objective was thus to avoid failing to be different.

A colleague of mine who teaches in the law school was involved with the creation of a new constitution for South

Africa. During the time that he was, he expressed to me an interest in the role of failure in engineering, evidently thinking that anticipating how a constitution might fail was a way of improving the chances for the new governing document to succeed. He subsequently informed me that when new governments come into existence, as they did in South Africa in 1983 and in Iraq in 2004, they have a need for "constitutional design," which is sometimes also termed "constitutional engineering."[5]

Nations and governments also need to have flags and other icons created. When the graphic artist Tariq Atrissi was asked to design a logo for Qatar, "to communicate to investors and tourists the country's ornate past and modern aspirations," he came up with a symbol that "combines blue for hospitality; gold for sand, sun and luxury; and burgundy, Qatar's national color."[6] Not all such assignments are so uncontroversial, and flag designers especially have to negotiate a minefield of ways in which they can fail.

Even before Iraq had a permanent government, a new flag was desired, and the design guidelines were essentially "to present Iraq as a Western country and to include references to the past." According to a contemporary description, "The new flag has a blue Islamic crescent on a white field and three stripes. Two stripes are blue, symbolizing the Tigris and Euphrates Rivers and the Sunni and Shiite branches of Islam, and the third is yellow, representing the Kurdish minority." Though the overall color scheme departed significantly from that used by most other Islamic countries in failing to contain the traditional "green (the prophet Muhammad's favorite), red as a symbol of Arab nationalism and white and black, referring to the battle standards of medieval Islamic dynasties," it was the blue used in the design that came in for special criticism. In spite of the designer's intended symbolism, critics believed the blue resembled that used by Israel, and so it was darkened. The flag's apolitical designer, Rifat Chaderchi, confessed that he had approached the problem as one of graphics and that he "didn't think about Israel." Thus, he made the fundamental error of not anticipating how his design could fail politically from the user's point of view.[7]

Chaderchi might not have made his mistake had he considered how other flag designs had failed or explicitly avoided failure. For example, to symbolize a postgenocidal era, the new Rwandan flag avoided the old one's "red and black, reminders of blood and mourning," using instead sky blue, yellow, and green.[8] Symbolism and imagery are real components of any design, not just that of the constitutions, logos, and flags of countries new and old.

Skyscrapers are highly symbolic structures, and failure to be sensitive to cultural influences has threatened otherwise successful monuments. Cesar Pelli, the architect of the Petronas Towers, was instructed to produce a Malaysian image for the monumental structures to be built in Kuala Lumpur. Since there were no indigenous structural models to follow, however, he looked to Islamic art for inspiration and came up with a floor plan based on a twelve-pointed star pattern. When he was informed that this was more Arabic than Malaysian symbolism, the pattern was modified to one featuring eight points connected with intermediate arcs. Had Pelli not already won a preliminary international design competition, he might have failed to receive the commission by his inattention to the smallest point. He later admitted that "the space between" the towers was his "greatest interest in this project." According to Pelli, that space was defined with the forms of the mirror-image towers and the "skybridge" that both links them and makes them into a "welcoming portal" that "symbolizes the threshold between the tangible and spiritual worlds."[9]

Many traditions, practices, habits, and even states of mind that might seem to be natural and without origin or conscious design were in fact introduced by inventive individuals and developed over many years by many others. Even in sports, where rigid rules would seem to militate against innovation, the failure of a player to score or the game to capture the imagination of spectators has driven invention.

The game of basketball was invented by James Naismith, a physical educator who faced a classic problem of design. In 1891, while working at the YMCA College in Springfield, Massachusetts, he wished to come up with a game that could be played indoors as an alternative to calisthenics and marching during New England winters. He wished to devise a game of skill rather than of brute strength. Among the other constraints of his problem were that the playing field, being indoors, had to be relatively small. He chose to use a readily available soccer ball and pair of peach baskets to devise an activity that was a variation on the childhood game of "duck-on-a-rock," which he had played outside a one-room schoolhouse in northeast Ontario, where he was born. Instead of trying to knock one rock (the "duck") off a larger rock by throwing still another rock at it, players of Naismith's new game had to sink the ball in the basket. He introduced the game with thirteen rules,[10] which naturally became refined over time. As in all sports, anything that was not explicitly against the rules was considered fair game.

Among the things the rules did not limit was the time that a team could hold the ball. University of North Carolina basketball coach Dean Smith exploited this "hoophole" and became famous for his "four corners" offense, in which players remained far from the basket and passed the ball among themselves to draw the defending team out away from the basket. In a game in 1979, the Tar Heels ran this offense for the entire first half, at the end of which the Duke Blue Devils led 7-0.[11] Three years later, in 1982, I saw the offense run with equal determination but more success by Maryland playing against Duke. The score remained close throughout a rather conventional first half but, the game being played in Duke's Cameron Indoor Stadium, Maryland remained the underdog. To take away the home-team advantage in the second half, Maryland's coach Lefty Driessel ordered his team into an extreme four-corners offense. The Duke Blue Devils did not go out after the Terrapins but simply remained in position around the basket, and so Maryland held the ball for most of the rest of the game, passing it back and forth in a rhythmic (and boring) manner until they wished to score. Maryland won 40–36 over a seemingly disorientated and virtually hypnotized Duke team.[12]

Lefty Driessel's Terrapins could play such a mesmerizing offensive game because at the time there was no rule to prevent it or the boredom that it produced in most spectators. Indeed, it was that kind of low-scoring, low-action game that contributed to the introduction of the shot clock in college basketball, an innovative rule change that made the game faster and higher scoring and so more exciting to watch. Whereas the old rules did not prohibit any team from doing what North Carolina or Maryland did, the desire to score had been so ingrained that not attempting to do so for almost twenty minutes evidently did not occur to most coaches.

Making sporting events more interesting and exciting to spectators is nothing new, but introducing a novel play or move takes a special act of invention and design. Today, the jump shot is part of virtually every basketball player's repertoire, but before the 1930s and 1940s it was unknown. The invention is variously credited to the "father" of the one-handed shot,[13] Hank Lusetti, who played for Notre Dame,[14] and Kenny Sailors of Wyoming,[15] who found that he was able to shoot over taller players if he jumped while he was shooting. The dunk shot, which so thrills crowds today, was not part of the game until 1946, when seven-foot-tall Bob Kurland, who played center for Oklahoma, not only conceived of trying but also succeeded in making one. The professional National Basketball Association welcomed dunking to enliven games, but the National Collegiate Athletic Association banned it until the 1976 season. Kurland also blocked shots, but this physical invention led to the refinement of rules prohibiting goaltending,[16] something Naismith anticipated in his original rules.[17] What other innovations in basketball shots or rules there might be in the future may not be obvious, but it is a sure bet that the game, like all games, will evolve.

In spite of what some fans may think, sports are not matters of life and death. But life and death themselves are, of course, and the practice of medicine revolves almost entirely upon the proper understanding of failure and the design of proactive or reactive schemes to avoid or reverse it. The desire to combat illness and stall or reverse aging is nothing new.

Robert Hooke, who is commonly remembered for having lost out in arguments with Isaac Newton over credit for theories about the nature of light and the law of gravity, was engaged in so many different areas of seventeenth-century science and engineering (including that related to the human body) that he has been referred to as England's Leonardo. Among his most famous discoveries is the one that is known as Hooke's law of elasticity, which states that a force exerted by a spring is proportional to its extension, a fundamental concept in structural engineering. In his own time, Hooke was most closely associated with the Royal Society, which was founded "for the promoting of Physico-Mathematical Experimental Learning,"[18] In 1662 he was appointed curator of the all-important experiments for the new society, whose motto was, "Take no man's word for it."[19]

Hooke was evidently a hypochondriac obsessed with what went into and came out of his body. And he was such a visceral experimentalist that he himself became an extension of the store of equipment for which he was curator. He was also an insomniac, and so he sought correlations between what he ingested before bedtime and how he slept. When he slept "pretty well and pleasantly" he reported dreaming of "riding and eating cream." He used himself as an experimental vessel and took a good many drugs and purgatives. He evidently believed that what has been referred to as his "pharmaceutical experimenting" enhanced and sharpened his mental ability.[20]

Robert Hooke was not the only member of the Royal Society who saw his own body as the ultimate curiosity and mystery of nature. Evidently many fellows made their bodies available for experimentation after death—the "Great Experiment." In one case, the results of an autopsy on a member who died of what was thought to have been a severe kidney stone were reported to a group of members at dinner. According to Hooke's diary entry on the case, it was "believed his opiates and some other medicines killd [*sic*] him, there being noe [*sic*] visible cause of his death."[21]

The range of visible observations had, of course, been extended in the seventeenth century by the microscope and telescope, with both of which Hooke was very familiar. Indeed, perhaps his most significant and enduring published work is *Micrographia*, a classic treatise on how to make and use the magnifying device. Among the illustrations in the book are one showing fine "gravel" in urine and the famous one of a flea, engraved on an 18 × 12 inch plate. Members of the Royal Society and like-minded natural philosophers were also interested in much more than studying insects under microscopes.

Human dissections, outlawed in the Middle Ages, had become public spectacles in the seventeenth century. However, as late as the nineteenth century physicians did not appreciate that the diseased organs that they handled during autopsies could be the source of harmful microorganisms that they transferred to healthy patients—especially women in labor—that they visited immediately afterwards.[22] By the twentieth century, dissection was done with more care and less openness, being confined mainly to the autopsy room and scientific laboratory. Now, when "the focus of medical science has shifted from whole organs to cells and molecules," and when "computerized scans and three-dimensional recreations of the human body provide cleaner, more colorful teaching tools than the time-consuming dissections of the past," the study of anatomy and the traditional dissection of cadavers are becoming things of the past. Not everyone in the medical profession approves of such a development, believing that hands-on dissection teaches a respect for the human body and the doctor's role in treating it,[23] not to mention giving future surgeons the opportunity to develop a feel for manipulating tissues and organs directly rather than through a computer mouse.[24] Dissection, especially in the context of an autopsy, is also an important link between success and failure in the practice of medicine.

Though death may be defined as the ultimate failure, not all failures are fatal. Heart failure occurs when that organ does not fully perform its function; when insufficient pumping action allows the lungs to fill up with fluid the diagnosis is congestive heart failure, but the prognosis can be optimistic. All medical diagnosis is in fact failure analysis. We go to the doctor when we sense that something is wrong with us, and we tell the doctor our symptoms with the expectation that they will be recognizable and with the hope that a definitive cure will be prescribed. The ability of physicians to diagnose symptoms and prescribe curative action is an apt metaphor for the role of failure in successful design. No amount of knowledge of normal (or successful) health suffices; doctors must know the pathological characteristics and causes of ill health (failure) to be able to design a successful cure. In fact, "Pathological anatomy is a field of inquiry that lies at the basis of all scientific medicine."[25]

But the practice of medicine is not all science. It also involves a lot of human judgment and provides plenty of opportunity for human error, including everything from administering a wrong dose to amputating a wrong limb. The mistakes range from innocuous goofs to fatal mistakes. How much error existed and should be expected and tolerated became the subject of much discussion in the 1990s, when an article on error in medicine was published in the *Journal of the American Medical Association* and the Institute of Medicine published its much-publicized report, *To Err Is Human*.[26] These frank appraisals of the situation elicited considerable soul-searching and calls for reform, but it was generally recognized to be unrealistic to presume that all mistakes could be eliminated. The best that might be hoped for was expressed by quoting Yogi Berra, who said, "I don't want to make the wrong mistake."[27]

Engineers have long appealed to the medical analogy in arguing for the thorough and systematic study of technical failures. According to the late-nineteenth-century American bridge engineer George Thomson,

> The subject of mechanical pathology is relatively as legitimate and important a study to the engineer as medical pathology is to the physician. While we expect the physician to be familiar with physiology, without pathology he would be of little use to his fellow-men, and it [is] as much within the province of the engineer to investigate causes, study symptoms, and find remedies for mechanical failures as it is "to direct the sources of power in nature for the use and convenience of man."[28]

The investigation and documentation of engineering failures is thus said "to contribute to improved design practices in the same way that medical pathologists have contributed to advances in medical science."[29] However, unlike physicians, who tend not to wish to admit their fallibility,[30] engineers are all too aware of it and of the possible consequences. According to one definition, "Structural engineering is the art and science of molding Materials we do not fully understand; into Shapes we cannot precisely analyze; to resist Forces we cannot accurately predict; all in such a way that the society at large is given no reason to suspect the extent of our ignorance."[31]

As base as the comparison may be, going to the doctor when we are ailing is not unlike taking our car to the garage when it is malfunctioning. In both cases, the professional listens to us describe the symptoms and poses hypotheses about the cause of the ailment. Before the introduction of computers, diagnosis depended largely upon a true physical examination, looking for abnormally functioning parts to fix. Now, increasingly, there is little manual examination, and many diagnoses are done largely by machines controlled by computers. Often, doctors are not authorized to perform procedures unless they are justified by the numbers generated not by human hands but by machines. The situation is not unlike that in the automobile garage, where the mechanic is not authorized to repair or replace a part unless told to do so by the "diagnostics."

Computers themselves work on the binary zero-one, on-off principle. Absolutely successful operation occurs only in the total absence of failure. A computer program perfect in every way but one may have a very mischievous bug, which can be as difficult to detect amid the lines of perfection as a tick is to see among the flowers in a field. Neither software nor hardware bugs necessarily cause a computer to crash. Indeed, a machine may operate perfectly fine with a bug, as long as it is not relevant to the operation at hand. For example, a system as conceptually simple as a lock and key can appear to be invincible as long as it used as designed. But the introduction of an unanticipated foreign object, even one of a decidedly lesser technology, can cause chaos.

Such a situation arose with bicycle locks of the kind that have a heavy U-shaped shackle that fits into an even heavier-looking cylinder, to which is attached a thick and heavy chain. The weight of the assembly presents special problems to bikers about how to transport it when it is not securing their wheels to a fence or other fixed object. To emphasize the strength of its locks, one company that made such a lock and chain is named Kryptonite, after the one substance that could defeat Superman. Still, it was somehow discovered by bike thieves that "mashing the empty barrel of a ballpoint pen into the cylindrical keyhole and turning it clockwise does the trick"[32] in seconds in opening the lock. Thus a plastic Bic pen that could be bought for pennies—if not just found lying about on the ground—was sufficient to defeat the purpose of the eighty-dollar security device. According to one critic, the lock company "was too slow to respond to the crisis, leading many customers to abandon the brand altogether." Kryptonite belatedly announced that "it planned to give its customers new locks at no charge," but this came only after videos were posted on Internet sites demonstrating the low-tech defeat of certain models.[33]

Even the absence of a hardware vulnerability does not mean that the operation of a system is without risk. Magnetic resonance imaging (MRI) testing is believed to be virtually perfectly safe, but under certain circumstances it can be deadly. Because of the strong electromagnet incorporated into the ten-ton machine, metal objects of all kinds are supposed to be absent from the examining room. Ignoring this procedural caveat has resulted in unfortunate accidents. A policeman saw his .45-caliber revolver pulled out of his hand and a round discharged. A woman who inadvertently left in a hairpin had it pulled up her nose and into her pharynx, from which it had to be surgically removed. Another woman, who had an aneurysm clip implanted in her brain, died when she underwent an MRI procedure. And a boy being tested was killed when an oxygen tank, which for unknown reasons had been brought into the room, was pulled into the center of the machine and fractured his skull.[34] Hospitals can be risky places, where seemingly the best-designed engines can surprise their engineers.

The Therac, which was developed in the 1970s by Atomic Energy of Canada, seemed to be the perfect medical device, for it could focus a beam of high-energy electrons on a tumor embedded deep in the body and destroy it without harming intermediate healthy tissue. Because of the high levels of radiation that could be generated by the machine, a variety of safety features were incorporated into the Therac-25 model to prevent overdoses. These newer models of the hugely successful device relied upon computer control to obviate such failures. Nevertheless, in the mid-1980s, several cases of fatal or severe radiation overdose resulted from use of the Therac-25, and the reason for the malfunction was a mystery.[35]

The software appeared to be error free. It regulated the dose of a prescribed treatment according to an operator's input of specified parameters. Among the parameters typed into the Therac-25 was an "x" to select the X-ray mode, which employed the machine's full 25 mega-electron-volt capacity, or an "e" to select the relatively low-power electron beam. Failure analyses discovered that on occasion the operator mistyped an "x" for an "e," but the mistake was caught before the beam was activated. Nevertheless, in some cases, even though the typographical error was evidently corrected before the activation, the higher dose was delivered, resulting in severe burns and, sometimes, a painful death.

Apparently what had been happening was that operators had become so familiar with the Therac-25 that they were able to use the keyboard extremely effectively. In fact, they could enter the beam activation order so quickly after changing an "x" to an "e" on the computer screen that the software had not yet processed the change, and so an X-ray rather than an electron beam was delivered. Patients so overdosed complained of feeling burned, but the operators were incredulous since the computer screen showed that the lower-energy beam had been delivered. It took a number of severe burns and fatalities before it was realized that being too quick in operating the keyboard did not allow enough time, even though it was measured in tenths of a second, for the machine to be reset before delivering the beam.[36]

No engineered thing or system is deliberately made inferior to its predecessors. But following models of success, especially without a historical context, more often than not leads eventually to failure. When a complex system succeeds, that success masks its proximity to failure. Imagine that the *Titanic* had not struck the iceberg on her maiden voyage. The example of that "unsinkable" ship would have emboldened success-based shipbuilders to model more and more and larger and larger ocean liners after her. Eventually, albeit by chance, the *Titanic*

or one of those derivative vessels would likely have encountered an iceberg—with obvious consequences. Thus, the failure
of the *Titanic* contributed much more to the design of safe
ocean liners than would have her success. That is the paradox
of engineering and design.

4

THINGS SMALL AND LARGE

Let us not take it for granted that life exists
more fully in what is commonly thought big
than in what is commonly thought small.

—Virginia Woolf[1]

In the nineteenth century the distinction between small and large things was more or less easily drawn. Small things were generally those that fit in the hand and that could be easily grasped, both physically and intellectually. They were typically simple things that—given the necessary materials, skill, time, and the simplest of tools—could have been crafted one by one by a single individual. Increasingly over the preceding century, they had also come to be made in quantity by division of labor. They were things like pins, needles, buttons, quills. They were small things easily forgotten in the history of technology.[2]

Medium-sized things were not so readily forgotten, but neither were they so obviously different—except in scale—from the large. These included the magic lantern, which was a marvel, as were the working models of steam engines and other mechanical devices that precocious and inquisitive children found fascinating. Many a budding engineer began by making his own models, learning about design in the process. The crafting of clocks and scientific instruments of all kinds was also a means of learning about how things worked, how they were made, how they failed, and how they could be improved.[3]

Large things were generally too big and heavy for a single individual to make, let alone lift, without the mechanical advantage of helpers and devices. These large things were the steam engines, steam boats, railroads, iron bridges, and other contrivances commonly used as symbols of the Industrial Revolution. Generally speaking, however, large things were scaled-up versions of and assemblages of smaller things. Though the whole was much greater than the sum of its parts, and the operating principles were not necessarily fully comprehendible by the uninitiated, they stemmed from the same creative source as did the smaller things.

As different in scale as small and large things could be, they were very similar in how they were conceived and designed. Thus, those who mastered the crafting of small things had a leg up in making the large. This was especially important in times when formal education was rare and virtually nonexistent for inventors and engineers. The historian Carolyn Cooper finds this idea embodied in "the myth of the Yankee Whittling Boy, which says that American inventions of the nineteenth century came from youthful practice with a pocketknife." She believes that the essence of the myth is captured in an "obscure poem published in 1857" by the Reverend John Pierpont,[4] which reads in part:

> His pocket-knife to the young whittler brings
> A growing knowledge of material things.

. . .

Thus, by his genius and his jack-knife driven,
Ere long he'll solve you any problem given;
Make any gimcrack, musical or mute,
A plow, a coach, an organ or a flute;
Make you a locomotive or a clock;
Cut a canal or build a floating dock;
Make anything, in short, for sea or shore,
From a child's rattle to a seventy-four;[5]
Make it, said I? Ay, when he undertakes it,
He'll make the thing and the machine that makes it.[6]

The underlying ability that could give a Yankee whittling boy—or any other inquisitive and industrious young person—the wherewithal to go from fashioning gewgaws out of twigs to building canals across America was the recognition of failure and its lessons. In making things small and large, it is always the perception and reality of failure that drives their invention, design, crafting, and evolution. For smaller things, the process had already spanned centuries, having brought many familiar objects to high levels of "perfection." For larger things, the limitations (failings) of hand-operated devices, such as the machines illustrated in Agricola's sixteenth-century monograph on mining,[7] had led to the development of the steam engine to lift water out of mines without the need of human- or horse-power. Watt's improvement on the Newcomen engine, whose cylinder had to be doused with water once each cycle to condense the steam, was motivated by its failure to use fuel and its heat economically. Watt improved the efficiency by employing a separate condenser.[8] But as much of an improvement as it was, Watt's engine still failed to achieve the full potential of its fuel, and so the engineering science of thermodynamics was invented and developed to better understand and guide how efficiency could further be increased. To this day, engineers look for ways to improve the efficiency of the engines they design.

The nineteenth century saw two different but interrelated developments that sharpened the distinction between small-ish and largish things. First, manual manufacturing practices, even with the advantage of division of labor, were failing to deliver in a reliable, efficient, and economic way a greater output of what we now call consumer goods. The introduction of machines to perform repetitive tasks automatically proved to be a boon to manufacturers. Not only did production increase and its costs decrease, but also the machine-made product was often more uniform in appearance and more predictable in performance. With a more reliably replicable product, the process of design evolution was accelerated. No longer could the poor performance of an item be blamed on the poorly made example at hand. Since each one was like every other one, a failure indicted not just the individual item but its design. The nineteenth-century patent literature is full of *improvements* in devices of all kinds and in the machines to make them.

If machine-made products were so much more predictable in form and function than their hand-wrought progenitors, then their combination into larger things also could yield a more predictable product. The development of the so-called American system of manufactures, in which interchangeable parts greatly reduced the time and effort needed to assemble or repair everything from rifles to magic lanterns to sewing machines, and especially Eli Whitney's role in it, continues to be debated by historians of technology.[9] But there can be no doubt about its contribution to the advancement of the Industrial Revolution.

Larger things like locomotives and bridges still had to rely on a great deal of manual labor to forge their parts and assemble them, but increasingly machines were developed at least to assist in doing everything from milling and boring to shaping and riveting. The fabrication and connection of the wrought-iron tubes of the Britannia Bridge, completed in 1850, required about 1.8 million rivets to be driven by hand.[10] The construction of the all-steel Forth Bridge, completed forty years later, benefited from the use of a machine to drive many of its 6.5 million rivets.[11] The overall length of the latter structure was almost four times that of the former, and their respective costs were £2.5 million versus £600,000.[12]

The second major development was the increasing assembly and construction of complex systems out of components. As large and critical as locomotives and bridges were, they were but minuscule parts of the system known as a railroad. And even individual railroads were dwarfed by the networks that had developed throughout Britain, America, and elsewhere by the later part of the century. There seemed to be no limit to the vision of engineers and businessmen because there is never any limit to their perceptions of failure. Even if all the continents were to be criss-crossed by rail networks, there would still be the disappointment that passengers could not cross the oceans by rail.

Isambard Kingdom Brunel had overcome that limitation

metaphorically. When confronted at a board meeting with misgivings about the magnitude of the proposed undertaking of extending the main line of his Great Western Railway much farther westward, he imagined continuing the passengers' journey to America aboard a steamboat named the *Great Western*.[13] Gustav Lindenthal, whose dream was to build a monumental bridge across the Hudson River at New York, thought in more tangible terms. He once remarked that "it was possible to bridge the Atlantic Ocean, but impossible to finance such an undertaking."[14] Bridging the Pacific Ocean might have been expected to have intimidated even engineers, but the proximity of Asia and North America across the Bering Strait captured the entrepreneurial imagination at least as early as 1890. That same year, on the occasion of the opening of the Forth Bridge, which carried the North British Railway across the last water barrier along the east Scottish coast, a souvenir program showed a locomotive named "Progress" pulling a through passenger carriage bearing the legend, "Aberdeen to New York, via Tay Bridge, Forth Bridge, Channel Tunnel, and Alaska."[15] Two years later, Joseph Strauss, who would go on to build the Golden Gate Bridge, proposed in his graduation thesis the construction of an international railroad bridge across the Bering Strait.[16]

The discoveries and advances in electrical science that took place in the early part of the nineteenth century inspired Samuel Morse to employ them in developing the telegraph.[17] The concept of communication at a distance was not new. Indeed, every time we speak with a neighbor across the back fence, we are communicating at a distance, albeit a very short one. The human voice can carry over far greater distances, as orators and yodelers have long known, and even unamplified it can reach large audiences. But there are clearly limits as to how far and how much content can be reliably transmitted. Visual means overcame these failings of the human voice, and in the late eighteenth century the French used semaphore-like devices to communicate across significant distances, relay stations being employed to pass the message on.[18] Such a system failed when visibility did, however, a limitation that Morse's idea of sending electrical signals over wires would not have. By the latter part of the nineteenth century, not only did telegraph wires net America and other countries, but also transoceanic cables interconnected continents.

The telegraph had its limitations, among them being the failure of individuals not versed in Morse code to be able to communicate directly with each other. Alexander Graham Bell's invention answered that challenge by allowing uncoded voices to be carried over unlimited distances, as long as they were connected by telephone wires. But the telephone too had its failings, chief among them being the very limitation shared by the telegraph: the need for wires. The development of wireless communication thus became the next great goal, and it was made practical around the turn of the century by, among others, Guglielmo Marconi. Since the "wireless," as the name suggests, was designed to replace the telegraph, it was perceived at first to be a failure in that it did not just carry signals point to point but broadcast them so that anyone with the proper receiving equipment could listen in on what were intended to be private communications. It took a while before this failing was seen as an advantage, but in a different context. Ships at sea could broadcast calls for help, as the *Titanic* tried to do. (Unfortunately, the nearest potential recipients of her distress calls had turned their wireless sets off for the night, not having yet come to recognize the full potential of the new technology.) It was not until after World War I that the advantage of broadcasting programs to what came to be called radios was fully realized and exploited.

In small and large things alike, in components as in systems, failure is a matter of perception. The telegraph was generally hailed as a marvel, if not a miracle. After all, "What hath God wrought?" The magical device that sent signals at lightning speed made possible what was not even a dream to ordinary people. But inventors are not ordinary people. If communication at a distance was limited to how far we could see and how clever we were at relaying messages, thought the ordinary person, then so it was. Perhaps it was part of God's design. But inventors also see every design as both a failure and a success, albeit one that necessarily has limitations—which they believe can be overcome.[19] They admire what a design accomplishes, but they see it as an old solution—and the basis for a new problem. A remarkable number of patents are followed up with subsequent improvement patents by the original inventors themselves.[20] To inventors, success is always transient, trailing failure behind itself like a fisherman trolling on a quiet lake. Inventors always take their own bait.

Inventors also swim in the sea of science and technology. They know its currents and they know its depth. They know what is possible today that was not possible yesterday. They connect old problems with new solutions, or at least with potential new solutions. Though some inventions may come fully formed in the inventor's mind's eye, they do not come fully developed in the hand. The conceptual design may come easily, but working out the details is generally plain hard work. As Edison put it, invention is 10 percent inspiration and 90 percent perspiration, and failure plays a central role in both.

Every new invention is presented as an improvement on the prior art. A patent application is a brief against the status quo. The form of a patent as literature tells the tale. As ostensibly different as patents for small and large things are, as different as a patent for a pencil is from one for a computer, there is a certain sameness to them. Typically, within the opening paragraphs of a patent application is a declaration, usually explicit but at least implied, of what is wrong with the prior art, how it fails to live up to its promise, how the inventor's solution addresses the problems, how it succeeds where the old failed. Thus, in his patent for an umbrella stand, William C. Carter stated, "In the umbrella-stands in general use, in which the umbrellas were placed in an ordinary haphazard manner, they were liable to become removed or stolen. The present invention relates to an improvement whereby this defect is remedied." Carter's 1885 improvement involved a rack onto which individual umbrellas could be neatly arranged and locked in place.[21] Simeon S. Post, in an 1863 patent for an invention of a different scale, declared, "The nature of my invention consists in constructing an iron bridge in such a manner as that the expansion and contraction of the material will not produce injurious effects upon the structure, and in this manner obviating one of the most serious objections to the universal use of such bridges."[22] An 1877 issue of *Scientific American*, which was then practically a gazette of patents issued, contained a column listing "new miscellaneous inventions" that consisted entirely of "improved" things, ranging from an "improved currycomb" to "improved combined overalls and jumper."[23]

As similar as is the fundamental cognitive and creative process by which small and large things come into existence and are improved, there are also significant differences in the consequences and implications of their failure in service. Familiar small things have generally been tested by years if not ages of use. The design of a new small thing, which has not yet been proven by the test of time, can be subjected to countless pullings, pushings, bendings, twistings, and flexings before it is even put on the market. Thus, prototypes of a novel design for a pair of glasses that employs a new material for the frame or a new spring mechanism for its temple pieces can be placed in a custom-made testing machine that simulates the spectacles being put on in the morning and taken off in the evening. The test can be automated to proceed without attention, the cycles of use even being counted automatically, and the machine set to stop whenever the glasses break—even if it is in the middle of the night and there is no one in the laboratory to witness the event. A technician, coming into work the next morning, can read and record the number of cycles to failure and load another prototype on the testing machine—gathering further data to make the tests statistically significant. Such tests provide statistical but not specific predictors of performance.

Not long ago, my wife was shopping for new glasses, and a rimless pair caught her eye. The temple pieces were attached directly to the plastic lenses by two small screws. When she asked the optician how well the design would hold up on someone who often fell asleep reading in bed, he said they were strong enough to survive being slept upon. To demonstrate their toughness, he pulled the temple pieces wide apart and flexed them with confidence—whereupon the right lens broke across the line of screw holes. He had performed the same demonstration on the same pair countless times before, he said, and this was the first time the glasses failed. He did not say whether his extreme test was a standard one in the factory laboratory.

No number of historic successes ever provides absolute insurance against future failure. Maker and user, let alone middleman, can have different expectations of what constitutes acceptable performance. Although manufacturers might like to reduce the number of reasonable and benign failures of a product to zero, it is as unrealistic a goal as obviating all product liability suits. Once mass-produced things have gone into production, quality control can be maintained by regularly sampling the output of the production line, but the creativity of users is beyond sampling. Dissatisfied customers are thus the source of important feedback on designs to manufacturers. Indeed, they provide the kind of information that is not readily obtained in any other way, for designers and manufacturers are often too close to a product to fully appreciate all the ways in which it can fail to live up to its promise. Designers and developers, working with a succession of prototypes, develop myopia as they learn to massage and manipulate the device to avoid its shortcomings. They define its limits and so test the product within them. Purchasers, however, seldom acknowledge such limits or read the fine print in the user's manual. No end of testing in a laboratory may ever uncover a flaw or weakness in a consumer product that can be immediately recognized by an imaginative maven.[24]

Larger and more expensive manufactured things tend to be more extensively tested by the company that makes them. The automobile industry is famous for its test tracks, around which still-to-be-unveiled new models are clandestinely put through their paces, perhaps simulating in months driving 100,000 miles. Airplane test pilots with the right stuff are legendary, as are their maneuvers subjecting new aircraft designs to conditions that are worse than the worst anticipated in service.

It is especially important to test complicated new physical systems, which can reveal behavior in use that was not anticipated in any designer's mind or in any computer simulation. Thus, the 1979 accident at the Three Mile Island nuclear power plant revealed unexpected ways in which instrument readings indicating valve positions could be misinterpreted.[25] A recalcitrant and misbehaving baggage handling system delayed the opening of the new Denver International Airport for over a year.[26] Computer simulations of such complex systems are themselves designed artifacts, and so they are subject to the same limitations of predictability and reliability as any other design.

Even space vehicles, such as those that carried the first men to the moon, undergo a regimen of testing. Such testing often takes the form of step-by-step demonstrations of proof of principle and proof of concept. In the early 1960s the whole idea of sending a human being into the weightlessness of space, much less going to the moon and back, was obviously new. Early demonstrations took place in the cargo bays of airplanes flying in parabolic trajectories to provide a moment of weightlessness. John Glenn, the first American to orbit the earth, went around only three times and spent less than five hours in space before reentering the atmosphere. Subsequent missions by other astronauts involved increasing numbers of orbits and days in space, showing that a critical component of the system—the human being—was capable of enduring the physical and psychological forces involved. Each flight also provided an opportunity to prove the workability of the recovery operation. The ultimate test could not be performed, however, in any way other than attempting the mission itself.

There is another kind of large thing that cannot be readily tested until it is fully built and tried. This is the civil engineering project—the dam, tunnel, building, bridge—whose scale is so large, whose cost is so great, and whose design is so specific to the site that the structure is unique. Because it is one of a kind, not made in a factory but constructed in place, there is no disposable example to test. Scale models may be employed for testing theories or comparing alternative designs, but no model will ever fully replicate conditions of the actual as-built structure. Even if incontrovertibly meaningful models were possible, it is not possible to simulate fully the natural forces of future earthquakes, wind storms, and the like to which the structure might be subjected. In short, the only way to test definitively a large civil engineering structure is to build it in anticipation of how nature will challenge it and then let nature take its course. This fact of large-scale engineering demands careful, proactive failure analyses.

That is not to say that every new large structure is a crapshoot. Engineers understand a great deal about the behavior and limitations of the materials and components of their structures. Steel, concrete, bolts, welds, and cables have all been tested extensively in a wide range of configurations and under a spectrum of conditions. Furthermore, it is the unusual structure that is built more than 10 or 20 percent larger than its predecessors (though there are some notable very successful exceptions, including the George Washington Bridge and the Empire State Building), thus moving only incrementally (and cautiously) into unknown territory. The bottom line is that no definitive test of the completed structure can be performed until the structure itself is complete.

There is a long tradition of "proof testing" completed large structures, especially bridges. Like the bridge designs themselves, such tests were prompted by prior failures. In the early nineteenth century, there were notable examples of suspension bridges collapsing under the feet of stampeding livestock, marching soldiers, and crowds of spectators watching spectacles on the water.[27] Subsequently, proof tests of new bridges and bridgelike structures naturally often took the form of soldiers crossing in step. A famous and very visible example, as reported by the *Illustrated London News*, was carried out in the presence of Queen Victoria when the safety of the galleries of the Crystal Palace was questioned.[28] A century and a half later, after the London Millennium Bridge was retrofitted with damping devices to counter the swaying motion that forced it to be shut down just three days after opening, employees of the engineering firm Ove Arup followed orders to move in synchrony to test the effectiveness of the modified bridge structure.[29]

Vibrations are imposed upon bridges also by mechanical traffic. Even the normal passing of trucks and cars can rattle the components of a bridge loose and to the breaking point, thus threatening its safety. But the first challenge to early railroad bridge designers was the sheer weight of a steam engine and its train. By the middle of the nineteenth century, it became standard practice to proof test new bridges by driving strings of locomotives onto them and noting the deflection of the structure. If the deflection was within the range expected under the load, the bridge was "proven." Such practices have continued, especially in Europe, where in 1994 the newly completed Pont de Normandie, then the longest cable-stayed bridge in the world, was subjected to the load of bumper-to-bumper tractor-trailer trucks filling its lanes to more than a third of the length of the main span. These dead-weight tests were supplemented by ones designed to simulate problems induced by vibrations. The center of the bridge was pulled down by a winch on a ship anchored in the Seine and then suddenly released to confirm that the bridge would vibrate as expected.[30]

In the late twentieth century, such physical proof tests fell out of favor in the United States. Rather than risk overloading a large structure in a proof test, so the argument goes, computer models of the structure could be tested to prove it. Like most arguments involving computer models, this makes sense as long as the structure as represented in the computer bears an accurate resemblance to the real structure. Strictly speaking, that is not ever likely fully to be the case, since artifacts of construction—flawed materials, loose bolts, forced connections, and the like—are typically not represented in the computer model. Thus, a computer-based proof test is not really one. Still, it is argued that computer models, which cost less to test, are suitable surrogates for real structures and systems.

Every large structure rests on a unique foundation and has its own idiosyncrasies of construction. Yet even a valid physical proof test provides no absolute insurance against failure. A proof test really proves only that the structure did not fail under the proof load at the proof time. The behavior of the structure under other kinds of loads, such as those imposed by traffic careening out of control, an earthquake, or moderately high winds—all of which should have been anticipated and designed against—will not have been tested. The hypothesis that the bridge will be able to withstand what designers anticipated could possibly cause it to fail can only be fully tested at the whim of nature. Like any scientific hypothesis, it can never be proved to the extent that a mathematical theorem can be; but it can be disproven ("falsified" in the language of Karl Popper[31]) by a single counterexample. In the engineering of large and small structures alike, that counterexample takes the form of an unambiguous failure.

Failure, not success, then, is the true touchstone of design. It would have been virtually impossible to have devised a proof test of the twin towers of the New York World Trade Center to withstand the combination of physical assault on its structure and the conflagration that ensued on September 11, 2001. That is not to say that structural engineers did not consider the possibility of an airplane crashing into the tall buildings. After all, a B-25 bomber did fly into the Empire State Building in 1945. Structural analysis of the effects of an impact of a Boeing 707, the largest airliner flying at the time, was carried out while the twin towers were being designed, and it was concluded that the structure could take such an insult without collapse.[32] In 2001, of course, Boeing 767s were deliberately crashed into the towers, which clearly initially were able to take the considerable structural damage that was inflicted upon them. This confirmed their mechanical robustness, but not their ability to withstand what followed.

The effects of the subsequent fires, which were initiated by burning aircraft fuel and fed by office furniture and paper, had not been tested against, however, in either a realistically meaningful computer simulation or a physically meaningful proof test. The former possibility, had it been carried out, may or may not have proven the structures capable of resisting such an inferno. Any computer test necessarily rests on assumptions about the conditions under which the assault on the structure will take place. In the case of the twin towers, a prescient designer of a computer simulation would have had to anticipate the extent of the structural damage, the destruction of fireproofing caused by the plane crash, and the amount of fuel and flammable materials present—not to mention wind conditions. In retrospect, this could have been done, but in prospect it would have entailed making in the late 1960s the combination of assumptions that in fact were to exist in 2001. The number of permutations and combinations of how a structure can be attacked is simply too great for a computer and its human handlers to test with a model of any sophistication. A fortiori, a physical proof test of the conditions that ensued on September 11 would have been unthinkable and probably ultimately meaningless. If a test of a completed tower involving the crash of a 707 and a fire had revealed weaknesses, they could have been fixed. Even then, the structure would not likely have been modified to the extent that would have been required for it to have survived an attack by a 767 flying at speeds far beyond what anyone would have considered reasonable for the relatively low altitude at which the towers were struck. If the tower did pass the 707 test, its design would have been "proven" safe, and nothing would have been changed.

Our built environment is populated by structures that have survived proof tests of all kinds. We term such structures successful designs, and they are. However, successful designs do not necessarily tell us very much about how close to failure they are. Sometimes cracks develop, which signal problems, but they can be attributed to settlement and interpreted to be the wrinkles of age. Any failure, however, is incontrovertible evidence that weaknesses existed—in the design, the workmanship, the materials, the maintenance, or the defense against terrorists. Failure is the counterexample to the hypothesis of success. This again is the paradox of design: Things that succeed teach us little beyond the fact that they have been successful; things that fail provide incontrovertible evidence that the limits of design have been exceeded. Emulating success risks failure; studying failure increases our chances of success. The simple principle that is seldom explicitly stated is that the most successful designs are based on the best and most complete assumptions about failure.

Human builders seem to have an irresistible urge to make larger and larger structures. For a long time, larger designs were modeled after smaller successful designs. Thus, pyramids grew geometrically taller, as did obelisks and Gothic cathedrals. Ships grew longer, as did bridges. As early as the seventeenth century, Galileo noted that there were limits to the geometric growth of structures, and he recited examples of those limits being marked by failures. Geometry by itself, he concluded, was an insufficient basis for enlarging a design. Natural structures, like the skeletons of animals, did not follow geometric rules, otherwise all animals should be expected to have the same proportions. Nature somehow incorporated something more into its designs. Galileo recognized that geometry had to be supplemented by considerations of the strength of materials.[33] And the key to his being able to unite the effects of geometry and strength in what engineers today would call design formulas was in Galileo's assumptions about how failure would occur. With his work, a new rationale for design was established, one that can accurately be termed proactive failure analysis.

5

BUILDING ON SUCCESS

Three things are to be looked to in a building:
that it stand on the right spot; that it be securely
founded; that it be successfully executed.

—Goethe[1]

What Goethe looked for in a building we should look for in any structure and, indeed, in any design. Even a needle must be pointed at the right place, must be held firmly in the fingers, and must be manipulated properly to achieve the desired sartorial end. Whether showing lantern slides or a PowerPoint presentation, the projector should be located in an appropriate place, it should sit firmly on a solid surface, and it should be properly operated. Certainly a speaker should prefer that the projector not be placed in a position where latecomers will walk in front of it, should not like to see it balanced on a wobbly stool or rickety rack, and should not want it operated by someone who cannot tell up from down or left from right, whether they be the sides of slides or the arrow keys of a computer.

The projection of slides is only one part of an illustrated lecture, of course. The design of the presentation itself should also conform to Goethe's expectations. We look for three things in a talk: that it be on target for the topic; that it be solidly grounded in fact and logic; and that it be effectively delivered. Not every effort at communication has to be so formally structured, but even among animals it often involves occupying the right territory, being firm in their claim, and asserting it convincingly. I have heard it often from cats, dogs, birds, and other animals in the neighborhood.

Having undermined our front lawn with tunnels, several chipmunks have set up residence in them. Their adits and exits dot the grass, and the cute little creatures scurry, hop, and bound about like, well, squirrels. In my imagined etymology, the name chipmunk comes from the high-pitched clicking sound that the ground squirrel is capable of uttering, sometimes with maddening repetition. I have come to recognize several of our resident chipmunks by, if not their voice, their territory and habits. One appears to own the middle of a line of three holes just to the right of the front door and pops in and out of it frequently, mostly in the morning. Especially when the grass needs cutting, this chipmunk often stands on its two back legs to survey the scene. Not infrequently, it mounts a rock on the edge of the lawn, presumably to get a better overview of its territory. A different chipmunk, one that is a bit larger and seems to reside somewhere around the side of the house, frequently sits on the top step of our front stoop and chips away, as if orating or running for office. From this vantage point, it commands an excellent view of the whole front lawn. It appears to be declaring ownership, and I have never seen another chipmunk challenge this.

Being atop a promontory is a desirable position for animals of all kinds, and one that we recognize as a position of opportunity, advantage, privilege, importance, power, and simple pride. As young children, we ride on the shoulders of our parents, thus enlisting both their mobility and their height to better move and see—and to be better seen. Climbing trees has an innate attraction for children, who seem to climb them just because they are there and they can. Fear of heights appears to me to be a learned phobia, one generally absent in the young. Being in a prominent position no doubt has its precedents in prehistory. Even the evolution of the erect posture of Homo sapiens may have been prefigured by the desire to have a higher and thus a better vantage. It not only allowed the hunter an improved shot at prey but also signaled dominance. Do not so many animals in conflict wish to make themselves appear to be as large and tall as they can?

Early shelters would seem to have been designed for retreating into and relaxing in rather than for posing and posturing in. As such, they need not have been high, generally speaking, and many peoples developed traditions of living on the ground. Where the climate demanded them, protective walls evolved as shelter from heat, cold, wind, rain, and snow. When settlements developed, at first they likely tended to spread horizontally. The construction of walled cities naturally limited the space on which to build, and so subsequent development within the walls naturally would have gone upward. There was an advantage to a loft or an upper story, where a house's occupants could sleep, as if in trees, above any household animals and out of easy reach of predators. If a loft were accessed by ladder, it could be hauled up; if stairs provided entrance to an upper story, their creaky steps could provide warning of intruders.

The pyramids are anomalies—assertions of height and

power whose raison d'être may still not be fully understood. They remained the tallest structures on earth for millennia. In medieval towns, the tallest thing was the cathedral, towering above the population like a lone tree surrounded by prairie grass. The Gothic cathedral showed that tall structures could be built to look light, but they were not projects to be undertaken without the blessing of a greater purpose, the resources of the faithful, and a proper respect for the possibility of failure. Cathedrals are also famous for seemingly being endlessly under construction and modification. One late-nineteenth-century discussion of building failures noted that the most common ones were those "due to injudicious alterations . . . in which the load carried by a building was greatly increased, owing to the idea that, as the structure and its foundations showed no signs of bursting, bulging, or cracking, there was a reserve of strength which might be drawn upon." Among the examples given were the "central towers of many cathedrals and churches," including Wells Cathedral, "where the progress of settlements due to raising a tower was successfully stopped by throwing inverted arches across the openings in the tower, so as to stiffen the work."[2]

During the Middle Ages, towers proliferated in northern and central Italian towns, proving the power of their owners and the prowess of their builders. But they were more for show and observation and signaling than for worship or shelter. And the successful completion of a tallest tower seldom failed to encourage an even taller one. In the nineteenth century building the tallest tower became an end in itself. When a competition was held for the reuse of the disassembled iron and glass components of the 1851 Great Exhibition building—the Crystal

Palace—one proposal was to erect a thousand-foot-tall Prospect Tower, whose design incorporated a forty-five-foot-diameter clock and which prefigured the skyscraper. Such an altitude was not actually approached for almost four decades, when in 1889 a tower of unprecedented height was completed for another international exposition. The wrought-iron Eiffel Tower owed its structural design to engineering experience with building railroad bridges (including such high viaducts as crossed wide and deep valleys like France's Garabit). The Eiffel Tower was the first iron structure to achieve the long-sought-after height of 300 meters (about 970 feet), which was twice as tall as the tallest pyramid. (The obelisk-like Washington Monument, which topped out three years earlier at 555 feet [185 meters], was the first structure in the United States to exceed the Great Pyramid in height. But neither the pyramids, the Eiffel Tower, nor the Washington Monument are considered buildings because they were not generally occupied in the normal sense of the word.)

Though bridge-building experience was a necessary prerequisite for designing the Eiffel Tower and, for that matter, the Crystal Palace, it was an ostensibly independent technology that made the more recent enterprise practical. It had been in another Crystal Palace, the one erected in New York in 1853, that Elisha Otis demonstrated his safety device for elevators, whereby the ever-present risk and fear of a hauling rope failing was rendered moot. With Otis's invention, if the rope did break, the elevator car would not plummet to the ground but by the action of a spring and ratchet mechanism would remain locked in place, its passengers safe and sound. That is not to say that early elevators were trouble-free. The same steam-driven elevator employed in constructing the Washington Monument was used to carry visitors to the top of the completed faux obelisk—in ten to twelve minutes. The elevator got so much use that it was inspected regularly by the Otis Brothers Company and by monument staff. Any sign of concern was cause to shut down the machine for repairs. Still, there was anxiety over its use.[3] On one occasion, the army officer in charge of public buildings and grounds reported to the chief of engineers, "It is believed that the elevator is as safe as it is possible for man to make it, and every effort is made to prevent accident; should an accident ever occur it will result from something which it was impossible to foresee."[4] Design is about nothing if not foreseeing.

And it was easy to foresee where the future lay. If a Washington Monument and an Eiffel Tower could be built as tall as they were, employing elevators to carry people to the top, so could commercially viable buildings intended to house offices and office workers. The new structural material steel, being stronger than wrought iron, could be employed to make lighter building frames, which could be built even higher. But before the widespread use of steel and the safety elevator, the height of buildings generally was limited by how many flights of stairs people were willing to climb. Elevators, even more than steel, are what made skyscrapers possible.

It should not be surprising that skyscrapers first took root in Chicago and New York City in the late nineteenth century. Each of them has water constraints, and desirable land to build on was not boundless in all directions. In New York, the first large buildings were erected in the vicinity of Wall Street, whose very name evokes the constraints of the city's early days.

In 1653, in anticipation of an attack on Manhattan from the New England colonists, Fort Amsterdam was repaired and a "high stockade and a small breastwork" was erected "across the town's northern frontier," the location of today's Wall Street.[5] In Chicago, which was also started as a fort, the downtown commercial district that was once encircled and constricted by trolley tracks is still referred to as the Loop, around which the elevated rail tracks now loop. It was out of such real and metaphoric constraints that the skyscraper grew.

As the 300-meter-tall tower was the goal in nineteenth-century Europe, so the 1,000-foot-tall skyscraper would become the goal in early twentieth-century America. Yet, as technically possible as it was to build ever taller, it was often difficult to justify on economic grounds alone. Most buildings were built as investments, with rental income providing the return on investment. But just building a skyscraper was no guarantee that renters would come. A prosperous firm might be able to justify building a skyscraper to house its own office operations, thus saving rent that it would otherwise have to pay, but few companies could fill a very tall building by themselves. Furthermore, an expensive skyscraper might cost much more per rentable square foot to build and operate than more efficient (and modest) office space. It was a business decision fraught with potential financial failure.

But there was an intangible benefit to building tall. Like an Italian tower, a prominent skyscraper signaled power and influence, the kinds of qualities most people wanted on their side. Thus, businesses like banks and insurance companies, which relied very much on image to distinguish themselves from the competition, were typically the ones that built skyscrapers named after the company. But building too tall too early might be counterproductive. After all, a 1,000-foot-tall skyscraper among 100- and 200-foot-tall buildings could not fail to be seen as extravagant and ostentatious, neither being a quality of a firm to which working people should want to trust their money or their life.

If banks and insurance companies could justify as sound the decision to build significant but not outlandish skyscrapers, then so could other businesses. The Woolworth Building was financed with cash derived from profits from the five-cent (later, five-and-ten-cent) stores of the same name. There had been growing opposition to building tall structures that blocked light from reaching the street and taxed the transportation facilities bringing people to and fro. But Frank Winfield Woolworth wanted a monument to himself as much as he did a profitable skyscraper. His chosen site, across from City Hall Park, was conveniently located near the seat of power in the city. It was also near the financial district and the Brooklyn Bridge, which made it further appealing to potential tenants.

When the architect Cass Gilbert asked Woolworth how high he wanted his tower to rise, he in turn asked how high it could be made. Gilbert responded that it was Woolworth's decision, and so the architect was instructed to make it fifty feet taller than the 700-foot-tall Metropolitan Life Tower, then the tallest building in the world.[6] The final height of the Woolworth Building, which opened in 1913, was 792 feet. The finishing touches on the Gothic structure that came to be called the "Cathedral of Commerce" included grotesques supporting the arcade galleries above the lobby. Among those involved in the building project who are carved in stone there are Woolworth himself, who holds a nickel; Cass Gilbert, who looks at a model of the building; the structural engineer Gunvald Aus, inspecting a beam; and Edward Hogan, "the building's rental agent, making a deal." According to Aus, it was Hogan who insisted on "a certain maximum and minimum size of office," a consideration that dictated the spacing between the steel framing columns. It was Hogan, also, who advertised the Woolworth Building as the "Highest, Safest, Most Perfectly Appointed Office Structure in the World, Fireproof Beyond Question, Elevators Accident-Proof."[7] The superlatives emphasize the competitive nature of building tall.

The role of ego (and whimsy)[8] in designing and commissioning structures cannot be underestimated. In Newark, Ohio, about thirty miles east of Columbus, there stands a seven-story building that is a replica of a market basket. The building, which is 160 times larger than the hardwood maplewoven basket after which it is modeled, serves as corporate headquarters of the Longaberger Company, the country's largest manufacturer of handmade baskets. The Longaberger Building, brainchild of David Longaberger, was topped with a pair of 155-foot long, 70-ton handles. Since the woven-basket design did not allow for normal fenestration, the building was fitted with a large but unobtrusive skylight over an open atrium. This in turn required a further design fillip: Because the building's basket handles, which arch eighty feet above the atrium skylight, would accumulate ice the way an airplane wing does in a winter storm, they are heated to prevent large chunks from forming and dropping onto the glass.[9]

Not all iconic buildings are as idiosyncratic as Longaberger's, but they can be just as much the product of an obsession, the most common being to be associated with the most distinctive or the tallest building in the world, at least for a brief moment in time. With the Woolworth Building being almost 800 feet tall, the 1,000-foot mark was no longer so much out of scale. By the late 1920s, three skyscrapers that would test the limits of ego would come to be on the drawing board virtually simultaneously. And it was not just their steelwork that commanded the attention of engineers.

Ironically, what limits the height of skyscrapers is not structural but mechanical. As buildings rise higher and higher, more and more elevators must be provided. Otherwise, the vertical transportation system will fail to move people quickly and efficiently, and so make the property unattractive to potential renters. But, if proportionately more space inside the building had to be devoted to elevator shafts, then proportionately less would be available to rent. Additionally, construction costs rise with height, and the rule of thumb would become that beyond about seventy-five stories, tall buildings were not financially viable. In other words, skyscrapers taller than that could be expected to fail to make a positive return on investment and so were generally not designed.

Then why is it that the Empire State Building, with its 102 stories, was ever built? The answer lay not in its absolute height but in the extraordinariness of that number of floors and of that height for the time. The project was announced in 1929, by former New York governor and recently defeated presidential candidate Al Smith, to be "the largest office building in the world and the largest single real estate undertaking in the history of the country." The enormous building would be "close to 1,000 feet high," he said,[10] but he would not be more specific.

By being so out of scale for its location on Fifth Avenue and Thirty-fourth Street, where there were no really tall structures in the vicinity, the usual rules did not apply to the Empire State Building. As Gustave Eiffel said of his tower when it was criticized by the Paris art establishment, "there is an attraction in the colossal, and a singular delight to which ordinary theories of art are scarcely applicable."[11] The Empire State Building was not only a colossal work of engineering art, it was the epitome of architectural design. And it was tall for the sake of being tall.

At the same time that the Empire State Building was being planned, the construction of two other enormous skyscrapers had already begun. One, at Forty-second Street and Lexington Avenue, was the Chrysler Building, initially advertised to be 809 feet tall and thus then besting the Woolworth Building as the tallest in the world. The other was the Manhattan Company Building at 40 Wall Street. When Walter Chrysler learned that, after revised plans, this building was expected to top out at 927 feet, his own building's target height was increased, though clandestinely.[12] Ultimately, Chrysler and William Van Alen, the architect of the great Art Deco structure, took everyone by surprise when the spire that had been assembled within the cloak of the building proper was jacked into place to give the completed skyscraper a total height of 1,046 feet. Thus the Chrysler Building stole the thunder not only from the Manhattan Company Building, which was supposed to reign briefly as the world's tallest, but also from the Empire State Building, which was supposed to be the first to exceed 1,000 feet when completed the following year. It eventually topped out at 1,250 feet (with the transmission tower and antenna added afterwards bringing the structure to its present overall height of 1,414 feet).

After this spate of tall buildings was completed in the early 1930s, none taller was constructed in New York until the twin towers of the World Trade Center were finished in the early 1970s. (After their collapse, the seventy-year-old prewar buildings returned to the position of being the tallest in the city.) The 110-story twin tower megastructures would not likely have been built by private enterprise, but the bistate, quasi-governmental Port of New York Authority was not driven by pure profit motives. Organizations have egos, too. Still, the decision to build not one but two such tall structures, containing a total of about ten million square feet of office space, had to be justified. Had the elevators needed to service the towers taken up more than the approximately 30 percent of floor area that they did, the towers might have had to be scaled down or reproportioned. However, an innovative scheme employing a system of express and local elevators—with transfers occurring in skylobbies—made the design work. By operating three elevators, one above the other, in each local shaft, a considerable amount of floor space was saved. The Sears Tower, which was completed just a year after the second World Trade Center tower and which at 1,450 feet recaptured the world's-tallest-building title for Chicago, employs a similar system of local and double-decker express elevators.

The Sears Tower reigned as record-holder for over two decades, until the Petronas Towers in Kuala Lumpur were completed in 1997. However, these Eastern twin towers were controversial for how they beat out the Sears Tower for the height crown. Like the Chrysler Building, though not so clandestinely, the Petronas Towers achieve their height by means of spires considered integral to the structure. The 1,483-foot-tall towers contain only eighty-eight occupiable floors.

The competition for the world's tallest building continued to capture the imagination of players and observers alike, but the attack on and collapse of the towers of the New York World Trade Center on September 11, 2001, changed the game. Suddenly, there was a new way in which a tall building could fail to be a viable design, and that was by being the now-credible target of terrorists. Supertall buildings financed and under construction generally went ahead as planned, but projects in the early planning stages were put on hold, scaled down, or scrapped altogether. This was the fate of the Chicago project designated 7 South Dearborn, named for the address of the building that would have won back the "tallest-building" title for Chicago and America. After September 11 Donald Trump scaled back his Chicago tower from 125 to 90 stories. Projects in the Middle East and East Asia, especially in locations perceived to be less susceptible or vulnerable to terrorist attack, generally went ahead as planned. Nevertheless, the chairman of the influential Council on Tall Buildings and Urban Habitat acknowledged that "the future of tall buildings seemed uncertain."[13]

In the United States, fear of terrorist attacks has affected not only support for future projects but also how existing buildings are perceived. In early 2004 a small professional firm moved into the Empire State Building, perhaps attracted by the post–September 11, 2001, availability of view-commanding space at attractive rates for the prestigious address. An open house to show off the new offices to old and potentially new clients attracted an embarrassingly small crowd. The view from the new offices may have been spectacular, but a lot of New Yorkers, images of 2001 no doubt still fresh in their mind, would rather not attend a party on a high floor in what they perceive to be a prime terrorist target. The open house was a harbinger of things to come, as the number of clients diminished and employees were laid off.

Herbert Muschamp, the architecture critic of the *New York Times*, closed his review of an exhibit on tall buildings mounted at the Museum of Modern Art by asking, "Does it need to be said that arousing fear is one of the things architecture is actually good for? Every visitor to the Eiffel Tower knows this. Tall buildings transport us to the far side of dread." Not everyone is acrophobic, of course, but fear of tall buildings has taken on new meaning since September 11. Muschamp also makes another remarkable statement in his review, namely, that "We assume that engineers can do wonderful things."[14] Indeed they can, but they can only do so much in allaying dread. Nevertheless, they can design structures that support tall buildings, and they can do so economically.

Even before the fear of further terrorist attacks on supertall buildings, their continued growth was limited by more than the elevator problem. As steel buildings grew taller, they naturally became more expensive to build. The structural frame had not only to hold up the massive weight of the building but also to stiffen it against the sideways push of the wind, which increased disproportionately with increased height. Roughly speaking, designing a building over about thirty stories tall meant incurring the additional costs associated with stiffening the structure against the wind.[15] This was the design penalty that Fazlur Khan, the structural engineer of Chicago's John Hancock and Sears towers, termed the "premium for height."[16]

It was in order to get a discount on this premium that Khan devised the tubular frame. Arguing that the most efficient structural form in resisting sideways bending was the tube—bamboo derives its structural strength from a lightweight structure that is hollow—he designed his tallest buildings following a tubular principle. Thus, the John Hancock, with its characteristic exposed diagonal steel bracing, has a tapered tubular frame. (The taper further mitigates the effect of the wind, while at the same time producing the desirable design byproduct of having the area of the commercial lower floors about twice that of the residential upper floors.[17]) The Sears Tower is nine square tubes bundled together, like a clutch of straws of different lengths, each one contributing to the strength of the group. The tubular form not only is efficient but also has the added advantage of pushing principal structural columns to the outside, thus removing obstructions from the interior space and making it more attractive to potential tenants.

The New York World Trade Center towers also employed a tubular principle. The closely spaced exterior columns gave the supertall structures sufficient stiffness against the wind. That is not to say that the towers could not be moved by the wind. Indeed, not knowing how occupants on the highest floors would react to the anticipated movement allowed by the flexibility of the towers, engineers designing the structures conducted tests on unsuspecting human subjects to see how sensitive they were to moving floors. A windowless test room was mounted on hydraulic actuators that could move the room to simulate what it would be like in an office at the top of the Trade Center. The test area was disguised as an optometrist's examination room, and subjects were led into it unsuspecting of its true function.[18] Learning in this way the tolerance of people to motion, that is, learning the amount of motion they would fail to accept, engineers could design the towers with the necessary stiffness.

Following its completion in 1976, Boston's tallest building, the sixty-story (788-foot-tall) John Hancock Tower, was infamous for having windows fall out for no apparent reason. The glass façade that reflected historic Trinity Church across Copley Square was patched with sheets of plywood while the cause of the façade failure was investigated. The principal problem was ultimately identified to be the manner in which the glass panels were manufactured, which made them susceptible to cracking and debonding from the spacers between the panes of insulating glass. In the course of seeking the cause of the window failures, it was found that the entire building, which had an unusual narrow trapezoidal footprint, moved excessively in the wind. To reduce the twisting motion, massive weights were connected via springs to the building's steel frame and tuned to its natural frequency. Such a system, employing what is known as a tuned-mass damper, is designed so that the weights can move contrary to any developing motion, thereby checking it the way a child can check the motion of a swing by moving contrary to it.[19] The troubles with the Boston Hancock tower were an early indication that buildings had become so efficiently designed structurally that they were overly flexible, sometimes in unexpected ways. Following and extending the

models of success provided by earlier buildings had pushed the
new structures closer to failure.

New York's Citicorp (now Citigroup) Center comprises a
915-foot-tall tower of unusual design. Its top is sliced off at a
45-degree angle to give a south-facing inclined plane reminiscent of rooftop solar-panel arrays popular at the time of the
building's construction, thus giving it a distinctive silhouette.
Completed in 1977, this skyscraper is also distinctive at street
level. The northwest corner of the parcel of land on which it is
built is owned by a small church, which did not wish to give up
its prime location. The church was willing to sell air rights
above it, however, and so the Citicorp tower was designed to
sit on four massive columns that are located not at the corners
of the building but at the midpoints of its sides. This makes for
dramatic overhangs of the tower's corners and for some unusual structural framing behind the façade. Because of the un-
conventional support of the tower, an extremely stiff and heavy
steel frame was required to keep wind-induced motion within
acceptable limits. The building was designed from the outset to
incorporate a tuned-mass damper, to avoid paying the full premium for height.[20]

Because of the Citicorp tower's unusual structural framing,
which had to direct the weight of the structure away from its
corners, the analysis of its behavior was not routine. In particular, predicting how it would react to wind forces involved analyzing the effects not only of winds hitting a face of the building
straight on, but also of "quartering winds," which are those that
strike the structure in a direction in line with a diagonal between two opposite corners. A frame with welded steel connections was determined to be sufficiently stiff to resist this

The distinctive base and crown of the Citicorp (now Citigroup) Center tower in New York City reflect the constraints and influences under which it was designed. (From the author's collection.)

"worst-case scenario," and the building's design was to incorporate such connections.

Shortly after the building was constructed, its principal structural engineer, William LeMessurier, discovered that somewhere along in the more detailed design process, bolted joints had been substituted for welded. Fearing that the effects of quartering winds had been checked only for a welded structure, LeMessurier recalculated the effects for the less stiff bolted connections and found the structure wanting in its ability to resist being toppled over. LeMessurier notified the building's owner of the potential disaster, the urgency to avert it being made all the more immediate because of the fast-approaching hurricane season. Rather than engage in litigious action, engineer and owner worked to retrofit the building with welded steel joints, which was done at nighttime, when offices were empty. For engineers, the building stands today as a testament to decisive action and cooperation and as a classic case study in ethical behavior.[21]

Though the imminent susceptibility of the bolted Citicorp tower to quartering winds was an acute problem, the chronic problem for tall structures of all kinds is moving too much in the prevailing winds. Thus, virtually all record-breaking skyscrapers being built today have motion-checking devices incorporated into their basic design. The Petronas Towers have long, heavy chains hanging inside their spires to mitigate unwanted motion of the buildings. Both active and passive damping devices are becoming an almost expected means of steadying tall, slender structures that otherwise would move in ways that would be uncomfortable and unsettling for some, if not most, occupants. The Taipei Financial Center, Taiwan's 1,667-foot-tall entry into the world's-tallest-building competition, is also known as Taipei 101 after its number of stories. It employs a large, heavy sphere suspended near the top of the structure. Rather than being hidden in some building cap or mechanical room, however, this enormous pendulum mass is a signature decorative centerpiece high above the streets of Taipei.[22] The building also has the world's fastest elevators (reaching speeds of 3,333 feet per minute to carry passengers from the lobby to the eighty-ninth floor in thirty-nine seconds). According to Taiwan's president, Chen Shui-bian, having the world's tallest building "not only gives affirmation to Taiwan's architectural industry, it's also the pride and honor of Taiwan's 23 million people."[23]

Claiming a world's record for height is not the only way to distinguish a structure. Another recently completed skyscraper is not so tall and slender that it needed mechanical damping devices to steady it the wind. Still, the overall shape of the 590-foot-tall building known by its London address, 30 St. Mary Axe, the street name referring to the weapon used in the fifth-century slaughter of virgins led by Attila the Hun, was designed in part to lessen the wind on the streets surrounding it. Because of the structure's unusual configuration, it has been described as having the "shape of a lingam"[24] and "variously compared to a cigar, a rocket, a bullet, a penis, a lipstick, a Zeppelin, a lava lamp, a bandaged finger, and—most frequently—a gherkin." However, its architect, Norman Foster, "prefers the metaphor of a pine cone or pineapple." An architecture critic finds "a transcendent sense of intellectual satisfaction," in the realization that the unique building is "more the result of tough, ingenious engineering and environmental solutions than the whims of clever, tasteful artists":

The Swiss Re Building, also known by its London address, 30 St. Mary Axe, is circular in plan. (Courtesy of Foster and Partners.)

The building is round in floor plan (every floor is a different size) to reduce the high winds generated at street level by tall rectangular buildings and minimize its apparent size: You cannot see the top from the bottom, and it tends to slip between surrounding buildings when seen from afar. The tapered lower half allowed the architect to open up a paved plaza in one of the densest parts of London; the rounded top softens its impact on the skyline.[25]

The building's "owner and chief tenant is Swiss Re, a sober and respectable Zurich-based reinsurance company" that is "seriously concerned about the possible financial costs to its clients of such things as global warming." Hence, it wanted its London headquarters to be "a model of 'green,' energy-efficient design." Among the green features of the building are "computerized external 'weather stations' that automatically monitor wind, sunlight and heat, and open or close windows and inter-pane blinds accordingly." Openable windows allow fresh outside air to "be guided about and used to reduce considerably the need for mechanical air conditioning."[26] In this regard, 30 St. Mary Axe is as much a machine as a building.

A ship is a transportation machine and also a long building that moves. Except when at anchor, it does not remain in one place, but at all times its captain must know its location on the chart—its latitude and longitude. The ship may not appear to be securely founded, but if its buoyancy is not properly ballasted it can topple over. Certainly the vessel must be successfully navigated, lest it end up grounded, or worse. But following an old channel through a harbor visited years earlier is no guarantee of successfully reaching the dock. Channels shift, and that is why there are harbor pilots. Rather than sailing around the world, they remain in one place, following not the stars but the shifting bed, over which only they can be trusted to pilot successfully.

Tradition, if nothing else, did not allow the Queen Mary 2, "the longest, tallest and widest vessel ever built," with its nearly skyscraper decks towering over many a low-rise building in the cities into which it sails with ultra-sophisticated instruments, to enter harbors without a local pilot. And although the Cunard Line "bragged" that its newest ship "could dock in many ports without tugs because of its state-of-the-art bow thrusters, controlled by joysticks on the bridge," tugboats were needed in New York because "tidal currents are very unpredictable in the Hudson." According to the ocean liner's master, "Sometimes it's slack on the surface, but a good current running 15 feet down." As successfully as the ship might have been maneuvered unaided in other harbors, such experience could not be relied upon in the new environment.[27] No matter how successful a design might appear to be, there is always the danger that latent failure is lurking beneath the surface.

6

STEPPING-STONES TO SUPER-SPANS

> Twenty men crossing a bridge,
> Into a village,
> Are twenty men crossing twenty bridges,
> Into twenty villages,
> Or one man
> Crossing a single bridge into a village.
>
> —WALLACE STEVENS[1]

Bridges, the greatest of which are among the largest and most ambitious deliberately designed structures in the world, have evolved more surely not in imitation of success but in response to failure. Colossal failures especially have been responsible for the most revolutionary changes in modern bridge design, but even the most primitive bridges developed out of responses to small and not so small annoyances, if not downright failures.

Fording a shallow stream has certainly always been a way to cross it, but the process necessitated getting one's feet, at least, wet in the process. Even less shallow streams could be crossed in this manner, but the swifter-flowing might have presented a challenge to the less strong and the less sure-footed. Deeper rivers could be crossed by swimming, of course, but at more risk and at the cost of getting even wetter. As natural as this might have been to our ancient ancestors and most animals, even a retriever wants to shake off the excess when it emerges from the water with a duck in its mouth.

The inventive mind makes connections out of diverse observations, and those who were stronger and taller might have volunteered to carry the weaker and shorter on their backs, perhaps for some spiritual consideration. Inventors are typically also of an entrepreneurial bent, and some may have seen considerable opportunity in devising ways of routinely carrying travelers and their burdens high and dry across chest-high waters. Extreme opportunists might have taken the traveler across first and gone back for the burden. Of course, the return crossing might never have occurred, the opportunists carrying the burden with them as they took off in the opposite direction across dry land. Improved and more secure ways of crossing rivers would have been especially welcomed by the weak, the trusting, and the gullible.

Inventive individuals of a different mind might have looked for an alternative to having to choose between getting wet and getting taken. When encountering a stream, such inventors might have walked along its bank, contemplating the problem. Eventually, a rapids might have been encountered. Though the water flowed faster and more forcefully around the rocks littering the bed, their fortuitous spacing might have suggested an alternative to getting wet at all. Even if stepping-stones reached only part way across the water, it should not have taken too much of a leap of the imagination to extend the naturally occurring path by relocating other stones to the desired place. Stepping-stones may thus have provided the first permanent dry crossings. Indeed, to this day bridges begin with very

widely spaced and high stepping-stones known as piers, between which the spans of the bridge are constructed.

A deeper stream would not have been suited to being
crossed by conventional stepping-stones, and this would have
challenged inventive travelers determined not to get their feet
wet. Another fortuitous placement of a natural object might
have provided an alternative. A fallen tree, its roots perhaps
having been undermined by the eroding action of the stream
itself, might have been tall enough to have reached from the
one bank to the other. Such a high and dry crossing might be
said to have been a found bridge, though one that required a
sense of balance to use.

Even had a fallen tree not reached entirely across the
stream, the inventive mind might have been inspired by the
situation to augment the partial bridge with a second log felled
from the opposite bank, or to fell another tree large enough to
span the stream all by itself. Such deliberately designed and
constructed crossings might be said to have been the first true
bridges. They served the intended purpose, but like all designed things they also had their limitations and failings. They
had narrow, rounded "walkways," and they were subject to rot
and hence to collapse. Though users of these bridges might
have adapted to them and accepted their limitations as being
in the nature of bridgeness, proto-engineers would have taken
the limitations as challenges to be overcome. Wider bridges
might have been constructed from two parallel felled logs, the
gap between them being spanned transversely by smaller ones
to provide a flat crossing, albeit one that had a ribbed roadway.
The material being timber, however, the problem of durability
would have remained.

A stone bridge made of flint was incorporated into the late-twentieth-century design of a pond at Stonecrop Gardens. (Photo by Catherine Petroski.)

An alternative material would have been stone, and in locations where it was available in relatively flat pieces it might have seemed the better way to build bridges. Piers could have been built up of smaller flat stones piled vertically and the piles spaced as far apart as the longest available rocks could reach. With longer pieces of stone spanning piers the way tree trunks did riverbanks, such beam bridges would have remained in place for some time. Indeed, stone bridges of indeterminate age still exist. A late-twentieth-century example made of a single piece of flint, measuring about fourteen feet long, up to four feet wide, and one foot deep, is in service in Stonecrop Gardens, located in the hills east of Cold Spring, New York, which is across the Hudson River from West Point.[2]

The nature of stone is that it is strong in compression but weak in tension. Stone used as a beam is tested in both modes, because its weight makes it sag and thus squeezes the top into compression and stretches the bottom in tension. Though ancient stone columns could be built tall, the spacing between them was limited by the strength of the lintels placed across their tops. Hence, Greek temples have closely spaced columns. Similarly, stone beam bridges had to have closely spaced piers, which meant that a long crossing involved the construction of a considerable number of them. Furthermore, the closely spaced piers would have constricted the flow of the river and so speeded it up between them, which in turn would have tended to scour the bases of the piers, causing them in time to collapse. It was to avoid this mode of failure that the Romans developed techniques for driving timber piles deep into a riverbed to provide firm foundations on which to build durable piers.

The natural limitations (read, *failings*) of stone beam bridges constrained their continued development toward longer spans. An alternative to building closely spaced piers was to construct corbeled arches out of longish flat stones. The corbeled arch is a structural form that each generation of children playing with blocks rediscovers. A distance greater than the length of a single block can be spanned by a stepped structure formed by stacking and counterbalancing blocks in an incrementally offset manner. Doing this from two relatively widely spaced piers results in a bridge. Some burial chambers in Egyptian pyramids appear to have been built in this fashion.[3] Corbeled arches also have their limitations, as children continue to learn over and over again. To build longer bridges with fewer piers required a change in structural form.

The origins of the semicircular Babylonian or Roman arch,[4] the "true arch" composed of truncated-wedge shaped blocks of stone, like the origins of prehistoric and ancient structural forms generally, can only be speculated upon. Wedges are among the oldest of machines. They were likely employed in splitting wood and quarrying stone for use in the construction of pyramids, obelisks, and other monumental structures. Aristotle considered the wedge in his "Mechanical Problems," asking "Why are great weights and bodies of considerable size split by a small wedge, and why does it exert great pressure?"[5]

In ancient times, the wedge would have been ubiquitous in quarries. In order for it to function, there first would have had to have been a natural or made crevice of some kind into which the wedge was inserted before it could be pounded upon. The wedge so placed was a bridge of sorts across the gap, and it was capable of bearing great pressure. Where a crevice was relatively wide, more than one wedge might have had to be used to span the distance. (The wedges, at least before being pounded upon, would have projected a distance up out of the crevice, this configuration alone might have suggested the design of an arch bridge.) After it had been driven sufficiently to split the stone, the wedge or wedges would not necessarily fall into the crevice, for the parts of the heavy stone need not have moved apart any great distance upon being split. Thus, the wedge or wedges would support themselves, bridgelike—across the chasm that had been created by their action—between the "abutments" of the split stone. The "great pressure" that the wedge exerted could easily have suggested that it itself could also continue to support great pressure, as it did during the pounding process. The splitting wedges in place thus provided a model for an arch bridge.

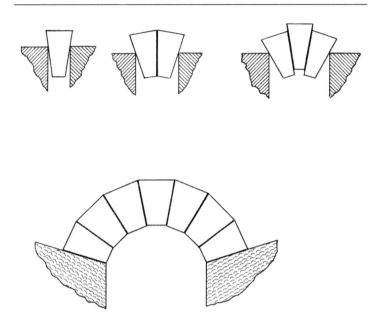

The use of splitting wedges may have provided inspiration for the invention of the stone arch. (Drawings by Charles Siple.)

A full-size arch bridge is formed by deliberately arranging blunt-wedge-shaped pieces of stone known as voussoirs between abutments that are massive enough to resist the great pressure created by the weight of the stones themselves and that which will bear down upon and cross over them. Of course, "deliberately arranging" the voussoirs is more easily said than done, because an incomplete arch will no more support itself than will a narrow wedge in a wide crevice. However, stone workers surely would have learned that it was helpful to have more than two hands to make a wedge bridge of more than two wedges. So it would have been obvious that what was needed to build a bridge of many voussoirs was a means of holding them in place until the last one (the keystone) could be inserted to complete the assembly of the arch.

The means of doing so was, of course, not stonecutter's hands but a stout timber scaffolding. This temporary structure—a bridge in its own right—came to be known as falsework, because it was not the true and final work. It also came to be known as centering, an especially appropriate term for the semicircular arch that became the ancient standard. The Roman arch bridge required heavy piers spaced no farther apart than the diameter of the semicircle. The width of the piers was typically about one-quarter to one-third of the arch's span,[6] a rule of thumb that no doubt developed in the wake of failures caused by testing the limits of those proportions.

One problem with the basic Roman arch was that its rise had to be at least one-half of its span, for that is the geometric nature of a circle. Thus, the longer the span, the higher the roadway atop the arch would have to be over what it crossed. This would not be a particular problem for an arch designed to cross a deep chasm or a river with high banks. However, where the banks of the river were generally relatively low, it would have been undesirable. Building a bridge that necessarily rose higher than the banks would require relatively long approaches to keep the grade of the roadway manageable. The higher the bridge rose, the longer the approaches would have to be, thus encroaching into the city streets and obstructing traffic traveling along the banks. To forestall such complications, a sufficient number of arches of modest span were typically employed in the bridge proper. (The Gothic arch, which in time became popular for aesthetic reasons, was seldom appropriate for a bridge because it emphasizes rise over span.)

A large number of short spans not only requires a large number of foundations and piers, thus increasing the cost of the bridge, but also obstructs the river and impedes navigation. Flatter arches, more elliptical in shape, enabled a greater span to be achieved without an accompanying higher rise, but they also exerted greater forces on the piers and abutments, requiring them to be thicker and heavier. This dilemma was resolved by the eighteenth-century French engineer Jean-Rodolphe Perronet, who argued from failure principles: Since the supporting piers had to be thick and heavy mainly to resist the pressures of a partially completed bridge during construction, which typically proceeded one arch at a time, a bridge whose arches were all built simultaneously would buttress themselves pairwise and so not require such stout piers. Perronet designed the flat arches of his Pont Neuilly to be built simultaneously and symmetrically, so that two adjoining arches would exert equal and opposite pressures on their common pier, thus eliminating the need for it to resist the push of an entire arch before the next was constructed. Thus the piers could be slender. The innovation necessarily required the trade-off that all falsework for the entire bridge had to be in place from start to finish, thus eliminating the cost-saving measure of reusing the timber as each arch was completed. Perronet's design also required that the entire falsework be removed at about the same time; otherwise, adjacent arches would exert unbalanced pressures on the slender piers, which might consequently be damaged. The striking of the falsework for the Pont Neuilly was a spectacle attended by crowds eager to see either history being made or a spectacular failure. Today, Paris is full of flat-arched bridges, making crossings between right and left bank appear to the pedestrian to be little more than extensions of the rues and avenues.

Where riverbanks were not low, semicircular arches continued to be employed for monumental structures. Stone bridges from Roman times, especially the spectacular high aqueducts across the Gard Valley in southern France and in Segovia, Spain, have survived for two millennia. They are obviously durable structures, but stone construction of any kind always had its drawbacks. Among the most significant were the time and cost involved, especially if there were not a convenient quarry from which to obtain suitable material. Where it was available, timber was an alternative to stone. A timber bridge could be built more easily and quickly, but it was more vulnerable to the ravages of spring freshets and, of course, subject to being destroyed by fire and rot. Fortunately, in the eighteenth century a third alternative became available for bridge building—cast iron.

The first iron bridge is generally believed to be the one-hundred-foot span completed in 1779 across the Severn River near Coalbrookdale, England, where an iron industry was flourishing. Keeping the river open to critical commercial traffic, while at the same time building a bridge to facilitate communication across the river, provided an opportunity for the iron founder Abraham Darby III, through his ability to cast large pieces, to show off the advantages of using a new structural material. As is the case generally with the use of a new material to make an old thing, proposals for the design of an iron bridge at Coalbrookdale followed structural and aesthetic models set by centuries of use of timber and stone. The bridge that was constructed resembles in its dominant semicircular form a Roman stone arch and in its connection details a timber structure. The promise of erecting the iron bridge without costly falsework, and thus without obstructing the river for the duration of construction, was a convincing reason to do so. How the heavy and unwieldy half arches of the bridge were erected had long been a matter of mystery and speculation, but recent scholarship has made a convincing argument that it was done easily with little more than muscle power and block and tackle.[7]

Cast iron was soon being used to make flatter-arch and, in time, beam bridges. But, like stone, cast iron is much stronger in compression than it is in tension. Indeed, the material's compression-to-tension strength ratio of about 6-to-1 dictated the cross-sectional shape of a cast-iron beam, with the area of its bottom flange being about six times that of its top.[8] Of course, generally speaking, the longer a beam is, the more it weighs. This in turn increases the stresses in the beam and so limits the span of a bridge that can be effectively made of it.

In the meantime, another type of iron became relatively widely available. Wrought iron is a much tougher material, and its strength in tension matches that in compression. This makes wrought iron suitable for making chains, rods, and other structural components better able to resist being pulled apart, and it makes the material ideal for use in suspension bridges. Although vine and rope suspension bridges are believed to have been made for ages by cultures in which their regular re-placement became ritualized, the use of iron chains for such bridges appears to have been employed first in China in the seventeenth century. The first Western iron-chain bridge is believed to have been built over the Tees River near Middleton, England, in 1741.[9] Applications of the suspension form accelerated with that of the increased production of wrought iron in the early nineteenth century. At that time, the longest spanning bridges came to be the iron-chain suspension type, and they not only were capable of carrying wider and heavier roadways but also had a longer serviceable lifetime. By the mid-1820s a suspension bridge with a span exceeding five hundred feet was built over the Menai Strait in northwestern Wales. Although cast-iron arches spanning six hundred feet had been proposed by the end of the eighteenth century,[10] one spanning more than five hundred feet would not be completed until 1874, when James Buchanan Eads employed the new material steel in his bridge across the Mississippi River at St. Louis.[11]

In the meantime, the longest simple-beam bridges were limited to being an order of magnitude shorter in span, with one estimate giving thirty feet as "just on the limit of cast iron" as late as the 1840s.[12] Then, it was the railroad that drove the development of bridges of many types. Though the cast-iron beam bridge was well suited to crossing relatively narrow canals, its limited span made it ill suited for direct use elsewhere on railroads. One popular means of designing longer bridges was to supplement a series of cast-iron beams connected in tandem with wrought-iron tie bars. This trussed-girder design was applied to longer and longer spans into the 1840s, but the 1847 failure of a nearly one-hundred-foot-long, trussed three-beam span of the Dee Bridge at Chester brought the extension of the form to an abrupt end. A century and a half after the disaster, one analysis attributed the failure to a crack initiating at "a sharp corner in a flange on a girder," a detail that "had presumably been added as an artistic flourish."[13]

At the time, suspension bridges were not believed to be suitable for railroad use, at least in Britain. Not only did the bridges lack stiffness, but also their roadways were prone to being destroyed in the wind, characteristics that were inconsistent with the needs and desires of the railroads. Thus, the pressing need for longer-span bridges had prompted innovative designs like the Britannia Bridge, which comprised wrought-iron tubes of such massive proportions that the rails were laid inside them. The individual spans approached five hundred feet, but the enormous cost of the bridge made it an uneconomical alternative to open trusswork designs, which came to be developed in great variety, especially in the United States. Of course, their spans were also limited. The suspension bridge could span much greater distances, and by midcentury it achieved free spans on the order of a thousand feet. John Roebling's Niagara Gorge bridge, the first suspension bridge capable of carrying railroad trains, spanned 821 feet in 1854.

The continued development of the railroad created a need for ever longer, stronger, and stiffer bridges to span increased distances. When a wider river, strait, or valley presented a new obstacle that familiar designs could not succeed in bridging, a new one would be sought. In the 1870s the expansion of the railroad up the east coast of Scotland produced some especially challenging problems. A coastal route between Edinburgh and Dundee had to cross the estuaries, known as firths, of the Forth and Tay rivers. The Firth of Tay was about two miles across at Dundee, but it was relatively shallow and so could be bridged within the state of the art with truss girders nowhere

exceeding 245 feet in span. The long, curving bridge was remarkable for being overall the longest in the world when completed in 1878, but in late 1879 its longest and highest girders
fell in a storm, killing the seventy-five people on the train
crossing it that night.[14]

The Tay Bridge's engineer, Thomas Bouch, had also been
contracted to build a bridge across the much deeper Firth of
Forth, and for its design he had chosen a tandem suspension
bridge with what would have been record-setting spans. It was
under construction when the Tay Bridge failed, and not surprisingly the commission for the Forth bridge was taken away from
Bouch. After an inquest and royal commission investigation asserted that the Tay Bridge had been inadequately designed, it
was rebuilt as a wider structure with more substantial-looking
piers. In many cases, the undamaged girders from the original
bridge were reused. The new bridge was built directly beside the
old, the stumps of whose piers remain visible today as a monument to the historic failure, which continues to be studied and
reappraised.[15]

The building of a redesigned Forth bridge was entrusted to
John Fowler, a mature engineer of impeccable credentials, who
had been so concerned about the Tay Bridge's stability that "he
had refused to let his family cross it."[16] Much of the responsibility for the Forth project went to Fowler's young assistant,
Benjamin Baker. It was he who would write, "If I were to pre-
tend that the designing and building of the Forth Bridge was
not a source of present and future anxiety to all concerned, no
engineer of experience would believe me. Where no precedent
exists, the successful engineer is he who makes the fewest mistakes."[17] Robert Stephenson, the engineer of the Britannia

A cantilever bridge under construction across the St. John River in New Brunswick demonstrated a structural principle distinctly different from that of a suspension bridge. (From *Scientific American*, 1885.)

Bridge, had expressed similar concerns. In a letter to its resident engineer, Edwin Clark, he wrote, "You will say I am always conjuring up awful difficulties & consequences—my answer to this is it is an important part of the duty of an engineer."[18]

In the wake of the Tay tragedy, the bridge across the Forth had not only to be a substantial design but to look like one to the lay people who were expected to entrust their life to the railroad crossing it. Fowler and Baker chose an uncommon design of unprecedented proportions. The type of bridge was a cantilever, which has roots in the corbeled arch but is capable of enormous spans. The spans of the Firth of Forth cantilever are in excess of 1,700 feet, which made it the greatest in the world when completed in 1890, having surpassed in achievement even that of the world-famous Brooklyn suspension bridge, completed in 1883. The Firth of Forth bridge, like the new Tay, not only was stronger than its predecessor design but, with its massive proportions and spread stance, also looked strong and stable against the wind, and it stands today as one of the great bridges of the world.[19]

The evolution of bridges, like that of every other made thing, proceeds through the identification and overcoming of failings and limitations—both real and perceived—in the state of the art. And things can be perceived to be over- as well as underdesigned. For all of the achievement that it was, the Forth bridge was considered an overly heavy structure for its length. Hence, it was natural for other engineers to want to design and build longer and lighter cantilevers. One was under construction in Canada in the early twentieth century, under the leadership of Theodore Cooper, an engineer who had had a long and distinguished career. The Quebec bridge was to span 1,800 feet across the St. Lawrence River, and it was well along in 1907 when it collapsed, claiming the lives of seventy-five construction workers. A royal commission found the design woefully lacking. Cooper and his design engineer, in their striving for economy and in their carelessness in carrying out a record-setting design without due respect for its magnitude, had made some fundamental errors.[20] The details of design errors are not what matter to the owners and potential users of bridges, however, and they tend to associate the entire type with failure.

The natural reaction to such a failure, as it was with the original Tay Bridge, is either to rebuild the structure in a noticeably strengthened form or to build another distinctly different type in its stead. In the case of the Quebec bridge, it was rebuilt as a cantilever, but one that bore little resemblance to the original. Where the first Quebec was slender, faired, and even almost lacey, the replacement was stout, angular, and austere. To this day, the Quebec stands as the longest cantilever bridge in the world, a distinction it has held now for almost ninety years. It is not that longer spans have not been designed and built; they have just not been built as cantilevers. This is because there were other types that could be made just as long—and much longer.

From the middle part of the nineteenth century, the principal long-span bridge type was the suspension bridge. After John Roebling had shown with his Niagara Gorge Suspension Bridge that the form could carry railroad trains without excessive deflection and with his Brooklyn Bridge that it could span distances in excess of 1,500 feet and remain steady in the wind, the suspension was a logical choice for long spans. (The cantilever did challenge it during the period between the building of the Forth bridge and the collapse of the Quebec, but that was only a brief period in the history of bridges.) In the half-century after the Brooklyn, the record span for suspension bridges more than doubled, from the 1,595 feet of the Brooklyn to the 3,500 feet of the George Washington Bridge, which was completed in 1931. By 1937 the Golden Gate Bridge had increased the record span to 4,200 feet, a distance between supports that neither the cantilever nor any other bridge type known could even approach.

If the Brooklyn Bridge had a fault, it was that it took a long time (fourteen years) to build and was expensive ($9 million, "exclusive of land damages, interest, and so forth").[21] Taking too much time and money is a failing that always attracts inventive alternative proposals. In the case of the Brooklyn Bridge, its masonry towers were seen as an artifact of nineteenth-century architecture, and the distinctive stay cables that Roebling incorporated into his bridges to give them stability against the wind were seen as redundant in the heavy bridges designed for twentieth-century rail and, later, motor traffic. Consequently, new suspension bridges were built with steel towers and without stay cables. Steel towers did enable bridges to be erected much more quickly and economically, without sacrificing strength, but the elimination of stay cables would prove to be a costly mistake.

By the end of the 1930s, suspension bridges were considered so efficient to build that major ones were being erected in remote and undeveloped areas whose traffic demands required no more than two-lane roadways. These long, narrow bridges were also designed with very shallow decks, in keeping with the aesthetic that had been introduced with the George Washington Bridge. But whereas the George Washington's extremely wide and heavy deck, which hung from necessarily massive cables, gave it an inertia that helped keep it steady in the wind, narrower and consequently lighter bridge decks invited the wind to cause them to vibrate, undulate, and eventually twist apart. In 1940, when twisting developed in the deck of the Tacoma Narrows Bridge, it collapsed in a matter of hours—a fate that was famously captured on film.[22]

The ill-fated Tacoma Narrows Bridge was characterized by an extremely narrow roadway. (From the author's collection.)

Normally in the evolution of a technology, such an unprecedented and highly visible and dramatic failure would mark a
dead end for the genre, as it did with the trussed girder of the
Dee and the cantilever of the Quebec bridge. However, in the
case of the suspension bridge, there was no alternative design
that could span such great distances. At the same time, there
remained numerous sites and communities that needed or
wanted long-span crossings. As a result, after a hiatus that co-
incided with World War II and the early postwar years,
suspension-bridge building resumed. The Tacoma Narrows
Bridge was rebuilt in 1950, but with a wider, four-lane roadway
and an exceedingly deep stiffening truss. The Mackinac
Bridge, completed in 1957, also had its deck stiffened by
massive trusswork, and in addition the roadway had an open
gridwork allowing the wind to pass through the structure.
The 4,260-foot-span Verrazano-Narrows Bridge, completed
in 1964, was constructed with a stiff double-deck roadway in
place from the start, even though traffic volume did not immediately require the lower deck, which remained unused for
years.[23]

The Verrazano-Narrows remains the longest span in America. However, since its completion, longer suspension bridges
have been built elsewhere. The first to take the record outside
America was England's Humber Bridge, which was completed
in 1981 with a span of 4,626 feet. At the turn of the millennium, the longest suspension bridge in the world was Japan's
Akashi Kaikyo Bridge. With a main span of 6,529 feet, it is
more than 50 percent longer than the Verrazano. Longer
bridges still are on the drawing board, most notably one to
cross the Messina Strait between Sicily and mainland Italy.

This span, which will exceed 10,000 feet, has at various times been proposed to be a pure suspension bridge and a combination suspension and cable-stayed structure.

The modern cable-stayed form developed in the wake of World War II, when decisions were being made about rebuilding the many European bridges that had been destroyed in the conflict. Where foundations and piers remained relatively undamaged, rebuilding directly on them was an economically attractive prospect. However, replicating the old bridge would have meant limiting its traffic-carrying capacity to prewar levels. In the meantime, traffic volume and intensity had grown, and so bridges with greater capacity were desirable. One way of achieving this with the old foundations was to make the replacement superstructure lighter but stronger, thereby allowing the bridge itself to carry heavier traffic without overloading the substructure. The cable-stayed bridge appeared to be the ideal form to achieve these ends.

As distinct from a suspension bridge, in which the roadway is suspended by (usually vertical) secondary cables from the large main cables that are slung over the towers and give the bridge type its characteristic profile, a cable-stayed bridge has its roadway supported directly by multiple cables emanating from the towers. Commonly called pylons, the towers of cable-stayed bridges are generally lighter in mass and form. Furthermore, because there are no main-suspension cables, the cable-stayed bridge needs no expensive anchorages and requires no on-site, time-consuming cable-spinning. The factory-made stay cables can be installed fully formed, and their multiplicity provides a structural redundancy that the suspension bridge generally lacks. Furthermore, because of the wide variety of

An artist's rendering of a cable-stayed bridge showed an option that the firm of Figg & Muller Engineers proposed for Florida's Sunshine Skyway. (From the author's collection.)

ways in which multiple cables can be arranged to connect roadway to pylon, the cable-stayed bridge provides many aesthetic options. Overall, cable-stayed designs are lighter, more quickly built, less expensive, and more individually distinct than suspension bridge designs. Not surprisingly, in many cases it has become the bridge type of choice, especially for communities seeking a so-called signature structure, as Boston did to carry some of the roads of its Big Dig over the Charles River.

When the cable-stayed bridge was introduced, the design was generally believed to be appropriate for moderate spans of up to about 1,200 feet. The classic suspension bridge design was thought to be the only one suitable for spans considerably longer. However, by the end of the twentieth century cable-stayed bridges with main spans on the order of 3,000 feet were being designed and built. (At the beginning of the new millennium, Japan's Tatara Bridge held the record at 2,920 feet, and longer spans were on the drawing board.)

Bridge design has come a long way from the placement of stepping-stones and felled logs across streams. Yet there remains a similarity of purpose and process between the imaginative ancient stone workers who saw a bridge in a span of wedges and the daring modern engineers who have seen farther by looking over the shoulders of their predecessors. The design process, whether it be applied to bridges or anything else, is timeless. It proceeds through persistence from failure to success. The failure of a single wedge to span a fault wider than it led to the use of multiple wedges. The failure of a single piece of stone to span a great distance led to the use of multiple spans. The failure of a single cast-iron beam to span greater than about thirty or forty feet (and only rarely as far as fifty feet[24]) led to the trussed girder. The failure of the trussed girder to span a hundred feet led to the wrought-iron tube and the open truss. Other limitations led to the use of suspension and cantilever bridges. The failure of the cantilever to withstand the rigors of construction led to its curtailment and to the dominance of the suspension bridge. And until the extension of the cable-stayed bridge into once unheard of realms, there appeared to be no alternative for long-span structures.

Conquering failure and overcoming limitations are what invention, engineering, and design are all about. Unfortunately, victory can be an intoxicant. Pride in the new successful thing typically turns in time to nonchalance, as the once-revolutionary thing becomes a common thing. Common things gradually become virtually invisible things, and the multiplication and extension of common things come to be done with little memory

of the failures of which they were born. When failure is left behind, success leads with a confidence that the uncharted future
does not warrant. In bridge building, following success means
taking a path that demands that ever-longer spans be based on
shorter versions. The path is fraught with danger and is epi-
cyclic. In time, success encounters failure again, and the pattern
repeats.

7

THE HISTORICAL FUTURE

There is not a fiercer hell than the failure in a great object.

—KEATS[1]

I was never afraid of failure;
for I would sooner fail than not be among the greatest.

—KEATS[2]

Design is Janus-faced, looking always both backwards and forwards. In the past, design sees at the same time an inspiring and yet an imperfect world, full of things to be both admired and improved upon. Of course, the past is also the repository of downright failures, monuments to ignorance, excessive optimism, and hubris. If heeded, the past thus provides caveats and lessons for future designs. If shunned, it will still haunt the future, always lurking in the shadows of success. In prospect, design too readily sees a world of perfection, one that is user-friendly and error-free. Under no circumstances should we expect that this will ever be universally the case.

Manned space missions present enormously complex design problems. In the Apollo program, not only did sophisticated hardware in the form of structurally and mechanically reliable spacecraft have to be designed, but also the software that controls and guides a mission had to be written and tested. To achieve the safety levels desired, redundancy was built into both software and hardware. Multiple computers were employed to check each other and to eliminate random errors. When the principle of the highly reliable Titan rocket, whose joints were sealed with a single O-ring, was adapted for the solid rocket booster of the space shuttle, a pair of O-rings was used in each joint. The presence of a second, backup O-ring ultimately allowed the shuttle *Challenger* to be cleared for launch even in the face of repeated evidence that the primary O-ring was not functioning as designed.[3] The fatal *Challenger* flight was the twenty-fifth in the series, which demonstrated a shuttle program success rate of 96 percent, far below the reliability it was believed to have. Prior to the accident, the "probability of failure with loss of vehicle and of human life" had been estimated to "range from 1 in 100 to 1 in 100,000," with the more pessimistic numbers coming from "working engineers, and the very low figures from management."[4]

After the obligatory hiatus, NASA resumed space shuttle missions with contrite caution. By the late 1990s, however, with *Challenger* a fading memory, the mantra of "faster, better, cheaper" was characterizing NASA's culture (though astute observers had often added "pick two," to emphasize that engineering and design always involve trade-offs). Notwithstanding embarrassing incidents associated with the myopic, if expensive, Hubble telescope and planetary probes that inexplicably crashed or were just plain lost (only one in three Mars missions was successful), NASA continued to forge ahead with confidence, if not swagger. In 1999 NASA suffered a plethora of embarrassments and failures, including the grounding of the shuttle fleet after the discovery of damaged wiring, the loss of the Mars Climate Orbiter due to confusion between metric and nonmetric data, more trouble with the Hubble, and the inexplicable loss of the Mars Polar Lander. According to Tony Spear, who was project manager for the 1997 Mars Pathfinder mission, which was "adequately financed and staffed," that and other contemporary successes had allowed NASA "to be too complacent in future project planning." Following "the first wave of success, the bar was raised for cutting costs even more, and it went too far," according to Spear.[5] At the same time, the management of programs had "grown too confident and too careless."[6] A policy analyst characterized the contrast between NASA's "faster, better, cheaper" philosophy and a more realistic one as "sloganeering versus engineering."[7]

After a while NASA seemed to have gotten back on track. But the destruction of *Columbia* in early 2003 brought things to an abrupt halt. It was once again clear that the space agency was subject to the same failing memory as aging individuals. Furthermore, just as some distinguished bridge builders have been guilty of hubris, so were NASA teams. Events leading up to the disintegration of the space shuttle *Columbia* on reentering the atmosphere constitute another classic example of success masking failure. Over the life of the shuttle program, insulating foam had broken off the external fuel tank on every launch, and the damage that it did had come to be accepted as a cost of flying. The piece of foam that came loose during the fatal launch of *Columbia* was the largest ever to strike the spacecraft, however, and it did so on a very vulnerable and critical part of the leading edge of the left wing. Though engineers

wanted to assess the degree of damage, NASA managers rejected requests to employ spy tools to inspect the orbiter in
space and dismissed concerns over its condition. "Over the
years, shuttle managers had treated each additional debris
strike not as evidence of failure that required immediate correction, but as proof that the shuttle could safely survive im-
pacts that violated its design specifications." With a mindset
similar to that which viewed O-ring damage and erosion before the loss of *Challenger* as not serious enough to delay its
1986 launch, again in the case of *Columbia*, "A generation of
NASA managers had turned engineering on its head, viewing
evidence of failure as signs of success."[8]

After the *Columbia* disaster, the shuttle fleet was grounded
for over two years. During that time, an accident board released its report and NASA worked to improve the design and
reliability of the remaining spacecraft, giving special attention
to reducing the amount and size of debris that could be shed
from the external tank during liftoff. NASA was optimistic
that great improvements had been made and that whatever risk
remained was acceptable, but not everyone agreed. In the
weeks before liftoff, there were accusations that NASA had
"loosened the standards for what constitutes acceptable risk."
Such debates over the reliability of a technology are not uncommon, for proponents and opponents alike are looking at a
concatenation of what-ifs and their probable but not certain
consequences. After some delays due to a faulty fuel gauge and
unfavorable weather conditions, *Discovery* was finally launched
in July 2005. Unfortunately, a sizable piece of foam—the kind
of debris that fatally damaged *Columbia*—came loose during
takeoff and the event was captured by the augmented system

of tracking cameras that had been put in place after the fatal flight. Fortunately, this piece of foam did not damage *Discovery*, but other problems that might have jeopardized reentry were dealt with in an unprecedented spacewalk. Even with *Discovery* still in orbit, and before any damage to it had been fully assessed, NASA announced that the entire shuttle fleet was once again grounded for the indefinite future. *Discovery* did return safely to Earth, of course, but the flight made evident how narrow a boundary can exist between success and failure.[9]

There are two approaches to any engineering or design problem: success-based and failure-based. Paradoxically, the latter is always far more likely to succeed. John Roebling, master of the suspension bridge form, looked not to successful examples of the state of the art but to historical failures for his education and guidance. From these he learned what forces and motions were the enemies of bridges, and he designed his own to resist those forces and suppress those motions. Such failure-based thinking ultimately gave us the Brooklyn Bridge with its signature diagonal cables, which Roebling meant to steady the structure against wind that he knew could be its undoing.[10]

Basing any design—whether of a product or a monument or a business process—on successful models would seem logically to give designers an advantage: They can pick and choose the best features of effective existing designs. Unfortunately, what makes things work is often hard to articulate and harder to extract from the design as a whole. Things work because they work in a particular configuration, at a particular scale, and in a particular context and culture. Trying to reverse engineer and cannibalize a successful system sacrifices the synergy of success and the success of synergy.[11] Thus, when bridge builders in the 1930s followed effective models that had evolved since Roebling's, they ended up with the Tacoma Narrows Bridge, at the time the third longest suspension bridge in the world and the largest ever to collapse in the wind. Over the years, in the process of improving on Roebling's design, the very cables that he included to obviate failure were left out in the interests of economy and aesthetics.

As a result of a freak accident, John Roebling died just before construction of the Brooklyn Bridge began. He was in his early sixties, and had circumstances been different he might have lived to design still another great bridge. It is interesting to speculate on what such a bridge might have looked like. His three principal suspension bridges were built at Niagara, Cincinnati, and Brooklyn. The first was steadied against the wind not only by stay cables stretching from the towers to the roadway but also by guy wires extending downward from the deck to the sides of the gorge. Lest they interfere with river traffic, below-deck guy wires were not included in the latter two bridges. Is it conceivable that there could have been a situation in which Roebling eliminated also the stay cables from some future design, perhaps to reduce costs or to satisfy a different aesthetic? Or, might his son Washington, as his father's intellectual successor, have succumbed to similar design simplifications?

The father's premature death and the son's deteriorated health make the question moot, but no major suspension bridge designer since then has incorporated the elder Roebling's stays, evidently believing them to be redundant, excessive, and unnecessary both structurally and aesthetically. For a century, the Williamsburg and Manhattan bridges, located just upriver from the Brooklyn, have withstood essentially the same weather conditions as it. But their stability in the wind no more proves the general uselessness of stay cables than did the Brooklyn's prove their usefulness, because all three bridges are at the same time similar and unique and, therefore, different. Only in the late 1930s, when suspension bridges with much more slender and narrow decks began to be built, did the value of the stays become evident. The decks of bridges designed by Othmar Ammann and David Steinman, each arguably Roebling's equal—or at least near equal—began to undulate in the wind. They were quickly retrofitted with a variety of supplementary cables and guy wires, as was the Tacoma Narrows Bridge. But even such measures were unable to keep it from eventually being destroyed by the wind, an incontrovertible vindication of Roebling's superior designs.

As described in the last chapter, the first Tacoma Narrows was only one in a historical series of large bridges that failed over the last century and a half. In each case, an instability developed in a bridge type that had theretofore been considered a success worth emulating with longer or more slender spans. Among the remarkable characteristics of this collection of failures, plus subsequent ones to steel box-girder structures, is that they occurred at intervals of approximately thirty years.[12] The pattern (1847, 1879, 1907, 1940, 1970) was so striking that it called out for the framing of the hypothesis that a major bridge failure would occur around the year 2000. In the early 1990s the most likely type to fail appeared to be the cable-stayed bridge,[13] a genre that might be thought of as having discarded the Brooklyn's suspension cables but retained its stay cables. In fact, the pattern was continued with a different class of bridges, in which the type of failure was characterized by the kind of loading rather than the structural principle involved.

The approach of a new millennium was the occasion for, among other things, a flurry of activity in designing long and distinctive footbridges, ones that were considered works of art as well as of engineering. Many were named Millennium Bridge, but it was the one in London that proved to get the most attention when it had to be closed only three days after opening because it moved so much under the crowds using it.[14] (The phenomenon was not unprecedented, and it had manifested itself most recently in a new footbridge across the Seine in Paris—the Passerelle Solferino, which had opened in late 1999.)[15] After being retrofitted with tuned mass dampers and other stabilizing devices, the London Millennium Bridge reopened in 2002. However, the still-young bridge is likely to be remembered for years to come as "old wobbly"—and as the bridge that continued the thirty-year pattern of failures.

Cable-stayed bridges remain likely candidates for a more dramatic failure, though perhaps now not until around the year 2030, given the heightened awareness of failure created by the Millennium Bridge and by problems associated with vibrations in cable-stayed bridges. Still, they are being built to lengths well beyond what was originally thought to be appropriate for their type. As if to signal their vulnerability to the wind, many cable-stayed bridges have exhibited unexpected cable motion, and they have been retrofitted with shock-absorber-like damping devices.

The ten-inch-diameter steel cables on a bridge across Dong Ting Lake in China's Hunan Province were swaying by as much as four feet in high winds. In an attempt to contain the motion of a representative cable, a magnetorheological damper was installed. This relatively new technology exploits the fact that a fluid containing iron particles stiffens in the presence of a magnetic field, thus reducing the amplitude of the motion. Such devices are "capable of sensing vibration in individual bridge cables and dissipating energy before it reaches destructive levels."[16] The demonstration damper worked so well on the Dong Ting Lake Bridge that all of its hundreds of cables were subsequently retrofitted with similar dampers.[17]

Increasingly, long-span cable-stayed bridges are having damping devices incorporated into their original design. Such is the case with China's Sutong Bridge, which will have a main span of 3,600 feet—almost 30 percent longer than that of the Tacoma Narrows Bridge—and will be fitted with magnetorheological dampers. Such measures may enable long spans to resist the wind, but they do not address the underlying problem of exactly why the cables vibrate in the first place. In fact, that is still incompletely understood, as was the aerodynamics of suspension bridge spans in 1940.

Another bridge type that appears to be evolving toward failure is the concrete box girder. This is a distant relative of mid-nineteenth-century bridges like the Britannia, through whose massive wrought-iron tubes railroad tracks ran. That wrought-iron form proved to be overly expensive and inappropriate for smoke-belching steam locomotives, but with the roadway relocated to the top of the box and with the material changed to steel, it experienced a renaissance in the middle of the twentieth century. In the 1960s steel box girders were being made in unprecedented lengths while the walls were growing disproportionately thin. With each successful application, the form "proved" its appropriateness to be pushed even further. In 1970 two steel box-girder bridges, one in Milford Haven, Wales, and another in Melbourne, Australia, failed while being erected. The tubes had grown relatively so slender and their walls so thin that they buckled.

In the case of the Melbourne West Gate Bridge, the erection technique used for the trapezoidal box sections led first to local buckling, which was corrected by loosening the box structure. This "method's success led to its repeated use when the next span proved to be far out of alignment," with eight-ton concrete blocks and jacks also being employed to deflect the box section into position so it could be joined to its mate. When some bolts were removed to achieve final alignment, the entire box buckled and fell on construction workers. According to the royal commission investigating the fatal accident, "Every feature of the West Gate Bridge had been used before, but not necessarily in combination or on such a large scale. Willingness to cope with imponderables is one thing that separates engineers from scientists, but it is imponderables that can provoke failures."[18]

Though steel box girders were indicted, contemporary concrete box-girder designs were not obviously subject to the same imponderables of plate buckling. By the end of the twentieth century, innovative methods of building concrete box girders of unprecedented length were becoming commonplace. A balanced cantilever method of construction was typically employed, in which the bridge is built incrementally out from both sides of its piers until the half spans met in the middle. With each successful application of the method, designers and

A post-tensioned concrete box-girder bridge across the Kennebec River at Bath, Maine, provides more clearance and traffic lanes than did the old narrow vertical-lift bridge, which now carries only the occasional railroad train. (Photo by Catherine Petroski.)

constructors appeared to treat it with increased familiarity and were emboldened to try longer and longer spans with seemingly less and less careful attention to details. In 1977 a record span of 790 feet was achieved in a post-tensioned concrete box-girder span constructed by the Army Corps of Engineers to connect the islands of Koror and Babeldaob in Palau, which was then part of the U.S. Trust Territory of the Pacific Islands. To correct progressive sagging of the structure, it was being retrofitted with additional post-tensioning tendons in 1996, when it suddenly collapsed.[19] It might be argued that the Koror-Babeldaob Bridge continued the thirty-year failure pattern, but the fact that it stood for almost twenty years puts it in a somewhat different category. That is not to say that newer concrete box-girder bridges do not bear watching.

Florida's Clearwater Memorial Causeway Bridge was designed as a concrete box-beam structure, and its construction by the balanced cantilever method began in 2002. Within a year an eighty-foot section of the bridge had settled and twisted. It was rebuilt, but other sections experienced problems of settling and cracking and had to be removed and rebuilt.[20] Though collectively these incidents were embarrassing to the designers and builders, they certainly could not be considered catastrophic. They did, however, appear to reinforce the impression that the concrete box-girder bridge type was not being respected for the massive structure that it is, at least in Florida. Though the Clearwater bridge cannot be considered a total failure, its construction troubles are certainly warnings to designers and contractors working on similar designs. In the short term, this renewed caution is likely to result in better-supervised concrete box-girder projects, but in the long term the vagaries of building in the form, and the attendant precursor and minor failures, may become accepted as commonplace. It may be just a matter of time before the lessons of Clearwater are forgotten entirely and the limits of stability are reached—perhaps around the year 2030.

Why have major bridge failures occurred at thirty-year intervals, and why should we expect the pattern to continue? The explanation offered by Paul Sibly and Alastair Walker, the researchers who first noted the phenomenon, is that thirty years is about the time that it takes one "generation" of engineers to supplant another within a technological culture comprising those working on a project or succession of related projects.[21] Though a new or rapidly evolving bridge type might be novel and challenging for its engineers to design, an older one that has become commonplace does not hold the same interest or command the same respect of a younger generation, who treat it as normal technology. Whereas the older engineers had an understanding of the assumptions and challenges that went into the basic design, as it and their careers evolved they went on to other things and so lost touch with it. At the same time, younger engineers, having inherited a successful design, developed no great respect for or fear of it. Thus, in the absence of oversight and guidance from those who knew the underlying ignorance, assumptions, and cautions best, the technology was pushed further and further without a full appreciation of its or its engineers' limitations. There developed "a communication gap between one generation of engineer and the next."[22]

A similar situation is believed to have prevailed at NASA at the time that *Columbia* was lost. The embarrassments of 1999 occurred thirty years after the spectacularly successful Apollo 11 mission that carried astronauts to the moon and back. In 2003 "at least a quarter of NASA's scientists and engineers were expected to retire within five years. Already the people over 60 outnumbered those under 30 by nearly 3 to 1."[23] And the older engineers had generally moved from engineering into management positions. A Caltech physicist characterized the situation as a "forgetting curve."[24] An institution might better remember how to avoid the pitfalls of large-project design by "keeping some people in a program from the earliest design stages until the end to maintain a continuity that could not be conveyed in written reports or computer databases," according to one space technologist.[25]

The problems are not unique to large civil and aerospace projects. The 1979 accident at Three Mile Island occurred three decades after construction began on the first peacetime nuclear reactor. The Y2K computer problem surfaced three decades after the programming languages FORTRAN and COBOL were employed widely in the 1960s. Indeed, many a retired computer programmer was called upon to bring to bear an expertise in a language that many in the younger generation of computer workers had not even learned. In late 1999 Japan's technological prowess was called into question following a year of embarrassing accidents involving the country's bullet train, its nuclear industry, and its satellite-launching capabilities. According to a professor of engineering at Tokyo University, the situation reflected "a lack of responsibility and an arrogance on the part of engineers and industry." The rash of failures signaled the end of thirty years of Japanese pride, if not blind faith, in the nation's extraordinary technological achievements.[26]

Research in literary history and in engineering would hardly seem to have much in common, but a recent development in the study of literature has revealed temporal patterns in the rise and fall of literary genres that are surprisingly similar to those related to the success and failure of large engineering structures like bridges.[27] (But perhaps the similarity should not be so surprising, for after all both bridges and books are designed things.) The literary scholar Franco Moretti has applied quantitative methods to the study of the novel, and in the first in a series of a projected three articles he proposed that the genre has had not a single rise but rather that different forms of it have developed, evolved, and disappeared in a repeating manner over the three centuries of its existence.[28] In fact, according to Moretti, the popularity of

different novel types—picaresque, gothic, domestic, etc.—
appears to have risen and fallen in roughly thirty-year periods,
indicating that there are forces at work that transcend any
given literary movement or fashion. As for why a thirty-year
cycle applies for the failure of styles of the novel, Moretti put
forth an explanation similar to that given for bridge failures: It
is generational, but not in a strictly biological sense. Rather,
the dynamic is more like the recurrent revolutions against
"normal science" that Thomas Kuhn so convincingly demonstrated.[29] Moretti posits that there is a "normal literature,"
manifested in a novel type, which gets replaced through some
process of intellectual destabilization. But he had no good
model for why such a "generational replacement" occurs so
regularly as every thirty or so years.[30] The same open question
essentially remains for bridge failures, which occur in an atmosphere of normal technology.

As different as have been the bridge types involved in establishing the thirty-year pattern, they all evolved according to
the general principles of design and the role played by success
and failure in the design process. The more commonplace
something becomes—whether it be the building of another familiar type of bridge span or the launching of yet another "routine" space shuttle mission or the writing of another novel in
the latest mode—the more there is a human tendency to gain
confidence from each success. The ever-present flaws and
glitches, and the little failures, become so familiar as to be ignorable, and they are ignored by all but the severest critics,
whose criticisms are usually also ignored or dismissed. Rather
than being seen as precursors to catastrophe, the little failures
are seen as nothing more than annoying blemishes attributable

to the imperfections of robust things and our understanding of them.

Even in the absence of a generational gap, individual engineers and designers can be susceptible to forgetting to be humble in the face of technology pushed toward its (unknown) limits. Robert Stephenson was one of the most distinguished engineers of the early railroad age. Yet it was he who was responsible for the ill-fated Dee Bridge, which in retrospect employed the trussed girder design beyond what was a prudent span length. Stephenson had used the type for many shorter spans, and its successful service in those applications evidently gave him an unwarranted confidence in the record-setting Dee, which collapsed in 1847. After the accident, the inspector general to the Board of Trade testified that, "he was opposed to this form of iron trussing and virtually admitted that he had approved the bridge because of the number of apparently successful ones already built."[31] According to the structural engineer James Sutherland, writing a century and a half after the fact, "It would be wrong to blame a single individual for the misconception over the behavior of these girders. This was a case of group myopia suffered by a large tranche of the most distinguished engineers of the day."[32]

Stephenson had been exonerated and went on to build the massive and daring Britannia Bridge that was an enormous structural success, albeit an economic and environmental failure.[33] He also became a staunch advocate of reporting failures, so that the entire profession might benefit from their lessons. According to Stephenson, who held the position of president of both the Institution of Mechanical Engineers and the Institution of Civil Engineers, "The older Engineers derived their most useful store of experience from the observations of those casualties which had occurred to their own and to other works, and it was most important that they should be faithfully recorded in the archives of the Institution."[34] Thus, engineers could gain the advantage of experience even without having had the experience directly. In effect, they could vicariously agree with the distinguished English biologist T. H. Huxley, who had said, "There is the greatest practical benefit in making a few failures early in life."[35]

Thomas Bouch had a successful early career. As a mature engineer, he believed that existing railroad bridges were over-designed, and he produced a Tay Bridge that when it opened in 1878 looked rickety and was rickety, collapsing as it did in the next year. Theodore Cooper was to cap a distinguished career in bridge design and analysis with the Quebec bridge, which was to be the longest cantilever in the world. Unfortunately, Cooper's failure to insist on the adequate supervision of engineers of limited experience, who were responsible for the design and construction of the structure, contributed to its collapse while still under construction, in 1907. Even David Steinman and Othmar Ammann, rivals who had each amassed distinguished records as world-class bridge builders, over-reached in the late 1930s in designing suspension bridges with overly slender roadways. The insufficiently stiff roadways of their Deer Isle and Bronx-Whitestone bridges, respectively, were excited by the wind into undulatory motion that required checking by supplementary cables. Each of these engineers could have been accused in his time of hubris, that arrogance that comes when the learning curve plateaus while the curve of ambition continues to rise. The result is the longer, lighter, leaner designs that grow out of an admiration of success rather than a full respect for failure.[36]

The problem of hubris and complacency in design was nothing new in the nineteenth and twentieth centuries and has not disappeared since. Some years after the collapse of the Quebec bridge and well before that of the Tacoma Narrows, the consulting engineer John A. L. Waddell, whose encyclopedic *Bridge Engineering* was published in 1916,[37] advocated the "independent checking of plans and specifications for all important engineering works by experts who have had nothing to do with their preparation."[38] Writing to *Engineering News-Record* in 1917, Waddell asserted that,

> If this independent checking of plans were practiced great disasters . . . could be avoided. It would add force to these claims were it possible to give a list of the most important failures of engineering works which have occurred during the past three decades; but so doing would raise animosities and cause hard feelings, which is something to be avoided whenever possible.
>
> . . . It is within the realm of possibility that at first there may be objections raised to this suggested innovation by some designing engineers; for they may feel that their rights and privileges are being encroached upon, their dignity offended, and their *amour propre* insulted.[39]

A surprisingly similar recommendation was made by Sibly at the end of his dissertation on structural failures.[40] At the same time he, like Waddell, recognized that the cost of independent checking could be significant. However, Waddell considered the cost "a bagatelle as compared with the economies effected by the avoidance of the expenses" of failures and the attendant loss of human lives. He also recognized that friction might develop between the checker and the checked, but he found this a small price to pay for the "security against disaster" that would result. For he felt it to be "almost inconceivable that two engineers, or two sets of engineers, working entirely independently of each other, should make the same error."[41]

But Waddell's recommendations remained generally unformalized, at least in the United States, and even when independent checking was carried out it did not necessarily have the desired effect. Securing federal funds required that a consulting engineer check the design for the Tacoma Narrows Bridge. However, Leon Moisseiff's self-assurance about suspension bridge design and his status as the leading theoretician of the form enabled him to rebut successfully the report of Theodore L. Condron, advisory engineer to the Reconstruction Finance Corporation, that the Tacoma Narrows Bridge was inadvisably narrow for its length.[42] Because Moisseiff had been involved with so many successful projects, he seemed not to be prepared for recognizing a potential failure in one of his own designs.

Over the past two centuries, suspension bridges have been especially and repeatedly prone to partial and complete failure, but the unique form has persisted nonetheless. The structural engineer Sir Alfred Pugsley, who worked on both aeronautical and civil structures, once "noted that about 1 in 14 suspension bridges had failed within their normal expected life span."[43] That amounted to an astonishingly low reliability of only about 93 percent. Things have improved considerably since Pugsley's survey. In the meantime, ever-longer suspension spans continue to be built, with the record holder of the early twenty-first century having a span well over twice as long as the 2,800-foot Tacoma Narrows. On the other hand the cantilever, once the bridge type that was presented as an alternative to the suspension for virtually every proposed long-span crossing, has for almost a century been limited by the redesigned and rebuilt 1,800-foot Quebec bridge. Why this inconsistent treatment of cantilever and suspension bridge types, the persistence of the latter appearing to contradict the general rule that a catastrophic collapse marks the end of a genre?

The answer appears to lie in part in the fact that the modern suspension bridge, dating from the earliest years of the nineteenth century, had a much longer history that the modern cantilever. The cantilever form was popularized in the 1880s, especially by the Forth bridge, only two decades before the Quebec tragedy, which was traced to something no more exotic than the bridge being too heavy to support itself. Ambitious experimental programs to test the ability of massive steel members to withstand great compressive forces ensued, but such quests for understanding seemed hardly modern science. After all, large static compression structures in the form of pyramids and stone columns had existed since ancient times. Although the modern suspension bridge form from its earliest days suffered numerous roadway failures, the cause given was repeatedly the mysterious and dynamic action of marching soldiers, gusting wind, and the like, phenomena that lent themselves to theoretical discussion if not definitive analysis. Advances in the theory of suspension bridge behavior, albeit under static conditions, provided ample justification for pushing the form into the realm of ever longer and more slender examples.

Immediately following the Tacoma Narrows collapse, it was evident that Roebling's sound if phenomenological analysis of a century earlier[44] had been largely forgotten in the half-century since the completion of the Brooklyn Bridge as "a triumph of intuitive engineering."[45] The so-called deflection theory promoted and employed by Moisseiff from the beginning of the twentieth century[46] had clearly been proven an insufficient successor. But more sophisticated analysis, incorporating mid-twentieth-century aerodynamics, provided ample justification for resuming suspension bridge building after the hiatus coinciding with the war.[47] Furthermore, suspension bridges longer than the Tacoma Narrows continued to stand, providing incontrovertible evidence that it was not absolute size, but relative proportion, that brought it down. Finally, with the curtailment of the cantilever, there was no alternative to the suspension bridge for crossing in a single span the long distances that remained to be spanned.

As bridge designs evolved toward longer and more flexible spans, so building designs have done so toward taller and more flexible towers. Though the unwanted motion of tall buildings appears to have been held in check, there may be reason for concern over the increasing dependence on tuned mass dampers and free-swinging pendulums to achieve this end. (Dampers of various kinds are being relied upon increasingly to mitigate the effects of earthquakes, being used in over 150 structures in the United States and over 2,000 in Japan, where they are required to be incorporated into "all major construction.")[48] Still, sky-scrapers appear to be evolving in ways akin to the evolution of suspension bridges in the 1930s. Taller and more slender super-tall buildings have been designed with regard for wind but not necessarily with full respect for it. By analogy with bridges, the response of ever more slender and flexible buildings to the wind is being fixed with ad-hoc patches rather than with a fundamental elimination of the effect. Bridge designers expressed no great urgency when new bridge after new bridge moved unexpectedly in the wind. Retrofitted cables were employed to check the motions while the problem was studied, but in the meantime more slender bridges continued to be designed and constructed according to the prevailing paradigm, which did not include a consideration of aerodynamics.

Tall buildings, at least in the East, are being designed in an analogous climate of complacency. The incorporation of damping devices is an acknowledgment that the underlying structure cannot handle the effects of the wind without assistance. The damping devices are expected to take up the slack, as it were. However, such devices have their own limitations and vulnerabilities. They involve moving parts, which are susceptible to developing cracks and failing, possibly at an inopportune time. They are also subject to being disabled, perhaps by some ill-intentioned person, thus rendering them ineffective. Most importantly, however, the limits of effectiveness of such devices may be in an unexpected mode of building vibration to which the damper is not tuned. The essential lesson of the Tacoma Narrows Bridge is not that it fell but that it fell in an atmosphere of confidence that it would not, and in a manner that was not anticipated. Constantly pushing the limits of experience of any technology is fraught with danger. It is done responsibly only a step at a time, and with a reality check after each that the wobble in the step is not getting out of control.

The collapse of a skyscraper was the stuff of fiction and motion-picture drama before the World Trade Center fell in 2001.[49] In his 1984 novel *Skyscraper*, Robert Byrne created a fictional building whose flaws did actually bring it down. (Since the discovery of the vulnerability of the Citicorp tower to the wind and the measures taken to decrease it had occurred during a prolonged New York City newspaper strike, its story was not widely known until it was the subject of a *New Yorker* magazine article in 1995.)[50] Through the early 1990s, the idea that such a thing could happen to a real building was believed to be literally the stuff of fiction. The outlandishness of the idea that a real skyscraper in a real city could topple was even reinforced after the failed attempt by terrorists in 1993, when their attack employing a truck bomb detonated in the public garage beneath the World Trade Center did not succeed. Although several floors of the basement garage were destroyed, the massive columns that held up the north tower remained intact. That the towers successfully endured such an assault was taken as evidence of their strength and stability. I recall, when visiting the World Trade Center in the wake of that attack, being struck by the noticeably increased security measures. These included a ban on public parking beneath the buildings and a bank of service counters in the lobby set up to clear visitors before they were allowed in the elevators. (Of course, they did not include defense against airliners used as weapons.) What I also recall from that visit in the mid-1990s was having not the slightest fear that the massive buildings were vulnerable to being brought down.

My suspicion is that, before September 11, 2001, few New York office workers or visitors to the World Trade Center or any other super-tall building had an inordinate fear of its collapse. Even the thought of a fire in a skyscraper was not especially worrisome. After all, fires that had occurred in tall office buildings had typically been confined to a single floor or two and burned themselves out or were brought under control without significant structural damage.[51] If there was anywhere to be concerned about a catastrophic fire, it was in a large hotel and not in a tall office building. The many lives lost in the MGM Grand Hotel fire in 1980 were a tragedy of a casino in far-away and frivolous Las Vegas, not of an office tower in close-to-home and businesslike New York.[52]

The terrorist attack of 2001 changed that perception irrevocably. Though people trapped above the impact floors futilely sought escape routes that were no longer there and suffered the unthinkable fate of being trapped more than a thousand feet above the street, those in the lower floors generally experienced considerable confusion but evidently did not panic. Nor did the emergency response workers, the police and firemen who valiantly entered the damaged structure and raced up stairs that others were racing down. By most reports, there was an orderliness in the confusion, and few involved seemed to fear that the collapse of the buildings was imminent. Being reminded of the fact that the twin towers had successfully withstood the truck bomb attack could only have reinforced that confidence. Now, of course, the unimaginable is vividly imaginable. No one can be accused of having an irrational fear of skyscraper collapse because it simply does not happen. The counterexample of the World Trade Center immediately diffuses any argument to the contrary. Failure is the most effective antidote to an inordinate belief in success.

Another example of pushing technology beyond its limits involves the former chief architect of Paris airports, Paul Andreu, "who stretches engineering to achieve his aesthetic goals" and "admits to having little interest in molding designs to structural purity." He has a reputation "for establishing dominance over the design team."[53] As chief architect, Andreu oversaw the expansion of the Charles de Gaulle airport, which involved designing and building "increasingly novel terminals." Terminal 2E was his last project before retiring from civil service and embarking on a second career designing "futuristic airport terminals around the world from Abu Dhabi to Singapore." The Paris structure was stunning, being an elliptical, tunnel-like concourse 100 feet wide and 2,100 feet long with no columns to impede the movement of passengers or block their view. The entire concrete-and-glass terminal building was supported on pylons and thus elevated off the ground so that airport service vehicles could pass beneath. The vast open interior space and the flattened curve of the exterior made for an architecturally striking building. Unfortunately, architecture alone does not support structures. About a year after the forward-looking terminal was opened, a 100-foot section collapsed with little warning, killing four passengers who did not flee in time.[54] The cause of the failure was not immediately apparent, but preliminary indications were that the concrete may have lacked some reinforcing steel critical to the bold architectural design.[55]

After a more through study of the failure, a government-appointed team found the overall design process to be at fault. The models employed to predict the behavior of the complex three-dimensional structure were much too simple to capture fully and predict accurately how the real one would work. Since it had been classified as a building, rather than as a civil engineering structure like a bridge, there was no requirement in France that the design undergo peer review. The failure, which occurred at a location where there was an opening for a footbridge on one side of the shell, was believed to be due either to excessive stress in some of the struts there or to the fracture of a beam opposite the opening. Even the way sunlight hit the structure may have played a significant role, heating and therefore expanding one side relative to the other. One member of the team analyzing the failure indicated that if the architect Andreu was to be faulted, it would be for thinking that the structure "was not so difficult to control." A contributing factor might also have been a tight budget, which did not allow sufficient money to be spent on the "very, very clever" structural designers that the complex project demanded. *Engineering News-Record*, generalizing about the situation, indicted a system in which "contractors of all sizes, seemingly addicted to low margins and high volume, live on the knife's edge of business success or failure with each project."[56]

Andreu, "who perhaps more than any other has set the standard for how airport terminals are designed,"[57] had wished to "prove his range went beyond airports"[58] and to be known as more than an airport architect. He has been described as "one of the most prolific foreign architects in China," having been responsible for a stadium in Guangzhou, the stunning Shanghai Opera House, and the newer National Grand Theater in Beijing. This titanium-and-glass structure, whose shape resembles that assumed by a water-filled balloon set down on a hard surface, has been variously referred to as a "glass bubble," as a "phosphorescent egg floating in a crystal sea," and, when covered with dust blown in from the desert, as "dried dung." Following the Paris airport terminal collapse, the national theater building came under increased attack, not only on aesthetic and economic grounds but also for possible political corruption in awarding the commission and for the safety risk seen in its underwater entrance.[59]

Daring architects and engineers have always walked a thin line between success and failure, and accepted the attendant risks. In spite of the failures they experienced in their careers, Robert Stephenson and Isambard Kingdom Brunel remain highly regarded in the pantheon of great Victorian engineers. The undulating bridges of David Steinman and Othmar Ammann were more or less forgotten among the great masterpieces in steel that they contributed before and after those embarrassments. Thomas Bouch and Theodore Cooper, on the other hand, had had only a body of pedestrian work to offset their misguided final daring bridge designs, and so their reputations have not fared nearly so well. After the Tacoma Narrows collapse, there was an attempt by his colleagues to protect the reputation of Leon Moisseiff, who as a generally low-profile consultant was an engineer's engineer both literally and figuratively. However, few other bridge designs were ultimately so directly connected with his name as was the Tacoma Narrows.[60]

In the wake of the collapse of the Paris airport terminal,

Engineering News-Record, editorializing about the construction industry, cautioned against retreating from daring designs:

> Unlike the manufacturing sector where factories often churn out millions or more widgets that are exactly the same, construction's curse and joy is that virtually every project is unique—made to suit a client's needs and matched to local conditions.
>
> Some clients like to push the envelope of earlier accomplishments. And some in the industry are probably not lying when they say that they can pretty much design and build anything, given sufficient time and money. And the lack of those two elements generally is where projects go sour. . . .
>
> There is no reason why owners should not take a walk on the wild side of design. That is how the industry moves ahead. But there should be no running, just to be safe.[61]

Daring structural designs are not only good for the industry. Without the likes of Paul Andreu and imaginative and innovative architects and engineers of the all kinds, our built environment would be sterile. In the worst imaginable case of over-conservatism, our cities would be built of identical structures, which would discourage the individuality of the people who occupied them. But no one wants to have to worry about unusual airport terminals, tunnels, bridges, skyscrapers, and other daring monuments collapsing over, under, and around them. And generally we do not so worry. It is precisely because catastrophic failures of large structures and systems are so rare that they are the news that they are.

Sir Alfred Pugsley wrote that, "A profession that never has accidents is unlikely to be serving its country efficiently."[62] This is not to say that engineers should be cavalier about safety, but that they must recognize that increasing reliability comes at a price. Since the cost of things like a manned space program or a monumental structure is determined not only by technical but also by economic and public policy goals, the reality of conflicts and trade-offs is always present. Risk and progress go hand in hand. Astronauts know and accept this, as do test and military pilots. Fighter planes can be made less prone to structural failure, but at the cost of weight and maneuverability, which would increase risk in combat situations. Risk and reward are relative terms.

The collapse of the Dee Bridge in 1847 led to the appointment of a royal commission to study the use of iron in railway bridges. The report ended,

> In conclusion, considering that the attention of engineers has been sufficiently awakened to the necessity of providing a superabundant strength in railway structures, and also considering the great importance of leaving the genius of scientific men unfettered for the development of a subject as yet so novel and so rapidly progressive as the construction of railways, we are of opinion that any legislative enactments with respect to the forms and proportions of the iron structures employed therein would be highly inexpedient.[63]

There is the tendency to impose restrictions in the wake of highly visible and tragic failures, but trying to legislate technological fixes is not always the most effective public policy for society as a whole. Because tile roofs had stood up well under Hurricane Andrew, which devastated southern Florida in 1992, many residential developments subsequently required their use in new construction. However, those same kinds of roofs experienced widespread failures in Hurricane Charley, whose winds "sent tiles like missiles into downwind neighbors."[64] Past success is no guarantee of future success.

Since September 11, 2001, there has been understandable interest in rewriting New York City's building codes[65] and mitigating the effects of fire in tall buildings.[66] However, requiring that all skyscrapers be designed to be able to withstand the kind of attacks that the World Trade Center towers suffered on that day might likely result in structures too expensive to build. Skyscrapers will always be vulnerable to extreme assaults. Improved fireproofing and fire fighting accessibility as well as incorporating more effective escape routes in designs are reasonably achievable goals, but they do not address the more fundamental question of whether such buildings will be considered safe by renters, office workers, and visitors. It is likely that for the foreseeable future in America, at least, highly visible skyscrapers generally will be rejected in favor of more modest buildings or low-rise buildings in nondescript industrial parks. Workers are likely to vote with their employment preferences, which will reflect their sense of security, narrowly focused though it might be.

If the goal of public policy is to be concerned equally for all citizens, then excessive attention to extremely low-probability events is not responsible government. It is not the daring and extraordinary structures and machines that should worry us but the routine and ordinary. In a typical year in the United

States, tens of thousands more people are injured and die in automobile accidents than in airplane crashes. Over twenty times more die in fires in one- and two-family dwellings than in nonresidential structures of all kinds.[67] And it has been estimated that "180,000 people die each year partly as a result of iatrogenic [i.e., physician-induced] injury, the equivalent of three jumbo-jet crashes every 2 days." Another study was reported to have found that there were "98,000 avoidable deaths a year that . . . might be caused by mistakes of doctors, nurses, and other hospital personnel."[68] A study of errors in an intensive-care unit found that they occurred at the rate of about one per one hundred "activities." Though medical workers might argue that, "Given the complex nature of medical practice and the multitude of interventions that each patient receives, a high error rate is perhaps not surprising," it is something we do not tolerate outside the hospital. Indeed, though a 1 percent failure rate might sound low, the corresponding 99 percent reliability rate is well below what we expect of other daily activities. According to one observer, "If we had to live with 99.9%, we would have: 2 unsafe plane landings per day at O'Hare, 16,000 pieces of lost mail every hour, 32,000 bank checks deducted from the wrong bank account every hour."[69] While technology can be and generally is remarkably error- and failure-free, there are no absolutes in engineering and design: "Systems that rely on error-free performance are doomed to fail."[70] Terrorist attacks may be an acute problem of the twenty-first century, but we should not ignore the more chronic risks that exist in everyday activities and situations.

Good design always takes failure into account and strives to minimize it. But designers are human beings first and as such are individually and collectively subject to all the failings of the species, including complacency, overconfidence, and unwarranted optimism. Given the faults of human nature, coupled with the complexity of the design of everything, from lectures to bridges, it behooves us to beware of the lure of success and to listen to the lessons of failure.

NOTES

CHAPTER 1: FROM PLATO'S CAVE TO POWERPOINT

1. C. H. Townsend, "The Misuse of Lantern Illustrations by Museum Lecturers," *Science* n.s., 35 (April 5, 1912), 529–531.

2. Simon Henry Gage and Henry Phelps Gage, *Optic Projection: Principles, Installations and Use of the Magic Lantern, Projection Microscope, Reflecting Lantern, Moving Picture Machine* (Ithaca, N.Y.: Comstock Publishing, 1914), 673.

3. Plato, *Republic*, Book VII.

4. David Hockney, *Secret Knowledge: Rediscovering the Lost Technique of the Old Masters* (New York: Viking Studio, 2001). See also David Hockney and Charles M. Falco, "Optical Insights into Renaissance Art," *Optics & Photonics News*, July 2000, 52; Lawrence Weschler, "The Looking Glass," *New Yorker*, January 31, 2000. But see also Sarah Boxer, "Computer People Reopen Art History Dispute," *New York Times*, August 26, 2004, E1, E5.

5. Susan E. Hough, "Writing on Walls," *American Scientist*, July–August 2004, 302–304.

6. Gage and Gage, *Optic Projection*, 674.

7. Ibid.

8. H. C. Bolton, "Notes on the History of the Magic Lantern," *Scientific American* 64 (1891), 277. See also Magic Lantern Society, http://www.magiclantern.org.uk/history1.htm. (Unless otherwise noted, Web sites cited were visited in 2004.)

9. Robert P. Spindler, "Windows to the American Past: Lantern Slides as Historic Evidence," *Visual Resources* 5 (1988), 1–2.

10. Bolton, "Notes on the History," 277.

11. Samuel Highley, "The Application of Photography to the Magic Lantern, Educationally Considered," *Journal of the Society of Arts* 11 (January 16, 1863), 142.

12. David Robinson, Stephen Herbert, and Richard Crangle, eds., *Encyclopaedia of the Magic Lantern* (London: Magic Lantern Society, 2001), s.v. "Huygens, Christiaan."

13. Howard B. Leighton, "The Days of Magic Lanterns," *Nineteenth Century* 5 (1979), 44–47.

14. "The Magic-Lantern," *The Manufacturer and Builder* 1 (1869), 199.

15. Robinson et al., eds., *Encyclopaedia*, s.v. "Huygens, Christiaan."

16. "The Magic Lantern," 199.

17. Leighton, "Days of Magic Lanterns," 45.

18. Gage and Gage, *Optic Projection*, 677.

19. "Lantern Slides: History and Manufacture," http://memory.loc.gov/ammem/award97/mhsdhtml/lanternhistory.html.

20. Spindler, "American Past," 13.

21. Leighton, "Days of Magic Lanterns," 46.

22. Townsend, "Misuse of Lantern Illustrations," 529.

23. Spindler, "American Past," 12.

24. Townsend, "Misuse of Lantern Illustrations," 530.

25. Robinson et al., eds., *Encyclopaedia*, s.v. "illustrated lectures."

26. Leighton, "Days of Magic Lanterns," 45. See also "Lantern Slides." For illustrations of the work of the Langenheim brothers, see George S. Layne, "The Langenheims of Philadelphia," *History of Photography* 11 (1987), 39–52. But see also Highley, "Application of Photography," 141, who believed "that to Messrs. Negretti and Zambra the honour is due of having first produced for public sale subjects of geographical and artistic interest, specially prepared for the lantern."

27. Simon H. Gage, "The Introduction of Photographic Transparencies as Lantern Slides," *Journal of the Royal Society of Arts* 59 (1911), 256.

28. "The Magic-Lantern," 199.

29. *The Magic Lantern* 1 (1874), 8.

30. Robinson et al., eds., *Encyclopaedia*, s.vv. "Marcy, L. J.," "sciopticon."

31. "Shadows in a New Light," *Chambers's Journal* 32 (July 12, 1859), 28–30.

32. Robinson et al., eds., *Encyclopaedia*, s.vv. "lever slides," "slipping slides."

33. *The Engineer*, quoted in ibid., s.v. "Constant, C."

34. Robinson et al., eds., *Encyclopaedia*, s.v. "Constant, C."

35. H. W. Whanshaw, *Shadow Play* (Darton, York: Wells Gardner, 1950), quoted in ibid.

36. Highley, "Application of Photography," 142–143.

37. Gage and Gage, *Optic Projection*, 200–201.

38. Robinson et al., eds., *Encyclopaedia*, s.v. "masks, slide."

39. Spindler, "American Past," 6.

40. Ibid., 7.

41. Robinson et al., eds., *Encyclopaedia*, s.v. "slide boxes."

42. Magic Lantern Society.

43. Robertson et al., eds., *Encyclopaedia*, s.vv. "Best Devices Co., Inc.," "cinemas, use of slides in."

44. C. Francis Jenkins, quoted in ibid., s.v. "cinematograph and lantern."

45. *Hope Reports Perspective* (Rochester, N.Y.) 1, no. 6 (July 1976), 1–2.

46. *Hope Reports* 2, no. 5 (May 1977), 1.

47. *Hope Reports* 1, no. 6 (July 1976), 1.

48. Ibid., 2.

49. Robertson et al., eds., *Encyclopaedia*, s.vv. "advertising slides," "Best Devices Co., Inc."

50. Magic Lantern Society.

51. Hope Reports, *Large Screen Presentation Systems* (Rochester, N.Y.: Hope Reports, 2000), 34.

52. See http://palimpsest.stanford.edu/byform/mailinglists/cdl/2003/1253.html.

53. Robinson et al., eds., *Encyclopaedia*, s.v. "Brown, Theodore."

54. Hope Reports, *Presentation Slides: Electronic and Film* (Rochester, N.Y.: Hope Reports, 1998), i.

55. The term "foils" continues to be used within the IBM organization, as I learned on a visit to the company's Yorktown Heights Research Center on November 7, 2004. In other circles overheads were known as "acetates" (see Stanley Bing, "Gone with the Wind," *Fortune*, May 2, 2005), 144.

56. Ian Parker, "Absolute PowerPoint," *New Yorker*, May 28, 2001, 76 ff.

57. See Henry Petroski, "Amory Lovins Guides the Hard Technologists," *Technology Review*, June–July 1980, 12–13.

58. Once, a computer scientist who was scheduled to give a presentation using an overhead projector in an auditorium with an especially high screen thought the keystone effect was distractingly strong. He solved the problem by redrawing his transparencies "with an anti-keystone correction, material at the top of each slide being much narrower than at the bottom." Peter Calingaert, e-mail to author, August 15, 2005.

59. *Hope Reports* 2, no. 5 (May 1977), 5.

60. Kenneth Silverman, *Lightning Man: The Accursed Life of Samuel F. B. Morse* (New York: Knopf, 2003), 156.

61. Parker, "Absolute PowerPoint," 80.

62. Hope Reports, *Presentation Slides*, 3, 5.

63. Ibid., 9.

64. Parker, "Absolute PowerPoint," 86.

65. CompuMaster, "How to Build Powerful PowerPoint Presentations," brochure for fall 2004.

66. Parker, "Absolute PowerPoint," 76.

67. See, e.g., Edward Tufte, *The Visual Display of Quantitative Infor-*

mation, 2d ed. (Cheshire, Conn.: Graphics Press, 2001); Tufte, *Envisioning Information* (Cheshire, Conn.: Graphics Press, 1990).

68. Ralph Caplan, *By Design: Why There Are No Locks on the Bathroom Doors in the Hotel Louis XIV and Other Object Lessons*, 2d ed. (New York: Fairchild Books, 2004), 237.

69. Edward Tufte, "PowerPoint Is Evil," *Wired*, issue 11.09 (September 2003). http://www.wired.com/wired/archive/11.09/ppt2_pr.html.

70. Edward Tufte, *The Cognitive Style of PowerPoint* (Cheshire, Conn.: Graphics Press, 2003).

71 Parker, "Absolute PowerPoint," 86.

72. John Winn, "Death by PowerPoint," *Journal of Professional Issues in Engineering Education and Practice*, July 2003, 115–118.

73. *New Yorker*, September 29, 2003, 97.

74. James Fallows, "Is Broadband Out of a Wall Socket the Next Big Thing?" *New York Times*, July 11, 2004, sect. 3, 5.

75. David Progue, "Office Buzz: Check the E-Mail," *New York Times*, September 25, 2003, G1, G4.

76. See, e.g., Charles Panati, *Panati's Extraordinary Origins of Everyday Things* (New York: Harper & Row, 1987), 106.

CHAPTER 2: SUCCESS AND FAILURE IN DESIGN

1. *The Encheiridion*, trans. W. A. Oldfather (Loeb Classical Library), 43. Quoted from Bartlett's *Familiar Quotations*, 16th ed. (Boston: Little Brown, 1992), 108.

2. Long, slender pointers would naturally be prone to breaking, and few of them survive in lecture halls today. Professor David Billington, who employs two slide projectors when he lectures to his class in an auditorium at Princeton University, makes use of a long window pole as a pointer. Throughout much of the lecture that I attended in December 2004, he rested the end of the pole on the stage.

3. Some years ago, as a memento of a lecture that I gave to the Graduate Materials Engineering Program at the University of Dayton, I was presented with a collapsible pointer of a handsome design. Closed, it is the size and weight of a slender Cross pen, but with a smooth, matte black finish. Telescoped open, the pointer betrays the chrome of an automobile antenna, but with a black plastic tip, which shows up well against the white background of the slides of equations, charts, and graphs typical of those then shown at a technical lecture.

4. Kenneth L. Carper, "Construction Pathology in the United States," *Structural Engineering International* 1 (1996), 57.

5. Jia-Rui Chong, "Whose Bright Idea Was This?" *Los Angeles Times*, February 21, 2004, B1.

6. Robin Pogrebin, "Gehry Would Blast Glare Off Los Angeles Showpiece," *New York Times*, December 2, 2004, B1, B9.

7. Damian Guevara, "Snow, Ice Slide Off New Building; CWRU Closes Sidewalk for Safety," *Cleveland Plain Dealer*, March 1, 2003, B1.

8. Michelle O'Donnell, "Rrrrrrrippp! Another Victim of Those Pesky Armrests," *New York Times*, May 28, 2004, B2. See also Terence Neilan, "Grabbing Long Island, by the Pants Pocket," *New York Times*, July 7, 2005, B2.

9. Michael Luo, "For Exercise in New York Futility, Push Button," *New York Times*, February 27, 2004, A1, A23.

10. Ibid.

11. "Plugging the Leak," *IEEE Spectrum*, January 2001, 81.

12. Katie Hafner, "Looking for the Eureka! Button," *New York Times*, June 24, 2004, G1, G7.

13. See, e.g., http://funnies.paco.to/mathJokes.html.

14. Armando Fox and David Patterson, "Self-Repairing Computers," *Scientific American*, June 2003, 54–61.

15. Ibid., 57.

16. "Striving for Dependability," sidebar, ibid., 60.

17. R. G. Elmendorf, letter to author, March 9, 2004.

18. Jessie Scanlon, "A Design Epiphany: Keep It Simple," *New York Times*, May 20, 2004, G5.

19. Fox and Patterson, "Self-Repairing Computers," 60, 56.

20. Linda Geppert, "Biology 101 on the Internet: Dissecting the Pentium Bug," *IEEE Spectrum*, February 1995, 16–17. See also "Pentium Pandemonium: What It Means for Engineers," *Civil Engineering*, February 1995, 26; "Intel Chips Away a Problem," *Engineering News-Record*, January 2/9, 1995, 10.

21. Peter H. Lewis, "The Inevitable: Death, Taxes, and Now Bugs," *New York Times*, March 7, 1995, C8.

22. Quoted in http://www.makingitclear.com/newsletters/newsletter4.html.

23. Juliet Chung, "For Some Beta Testers, It's about Buzz, Not Bugs," *New York Times*, July 22, 2004, G1, G7.

24. Quoted in "Ralph Baer: Video Games Wizard," *Prototype*, Newsletter of the Lemelson Center for the Study of Invention and Innovation, National Museum of American History, Smithsonian Institution, Summer 2004, 3.

25. T. J. Fogarty, "Embolectomy Catheter," U.S. Patent No. 3,435,826.

26. Jim Quinn, "'Failure Is the Preamble to Success,'" *American Heritage of Invention & Technology*, Winter 2004, 60–63.

27. Ibid., 60.

28. Dennis Boyle, quoted in Bart Eisenberg, "Thinking in Prototypes," *Product Research and Development*, January 2004, 28.

29. Quoted at http://www.bartleby.com/100/484.html.

30. Quoted in *Bartlett's*, 16th ed., 486.

31. Caplan, *By Design*, 229, 230.

32. Jeff Meade, "Profile: Jack Matson," *ASEE Prism*, October 1992, 36–37.

33. Michael R. Bailey, ed., *Robert Stephenson—The Eminent Engineer* (Aldershot, Hants.: Ashgate, 2003), xxiii.

34. See David P. Billington, *The Innovators: The Engineering Pioneers Who Made America Modern* (New York: Wiley, 1996), 105–108.

35. Brian Richardson, "Luas Continuing, Not Creating, History," *Irish Engineers Journal* 58 (2004), 362. For a table of principal world gauges see Douglas J. Puffert, "Path Dependence in Spatial Networks: The Standardization of Railway Track Gauges," www.vwl.uni-muenchen.de/ls_komlos/spatial1.pdf (March 16, 2005).

36. See, e.g., Frank Romano, "Technology, the Future, and the Horse's Behind," *Ties*, October 2000, 1.

37. George M. C. Fisher, "A 21st Century Renaissance," *The Bridge*, Fall/Winter 2000, 11–13.

38. See http://www.aleve.com/info_central/faq/faq.htm.

39. Del M. Thornock et al., "Package Exhibiting Improved Child Resistance Without Significantly Impeding Access by Adults," U.S. Patent No. 4,948,002.

40. Del M. Thornock and James R. Goldberg, "Bottle," U.S. Patent No. Des. 330,677.

41. Thornock et al., "Package Exhibiting Improved Child Resistance."

42. According to a consumer relations representative, only certain-size packages of Aleve caplets, tablets, and gelcaps come with the Easy Open Arthritis Cap. "Bayer Email" to author, October 20, 2004.

43. Quotes from a box and bottle of 100 caplets of Aleve purchased summer 2004.

44. Thornock et al., "Package Exhibiting Improved Child Resistance."

45. See frequently asked questions at www.aleve.com.

46. Quoted from box purchased summer 2004.

47. Package purchased fall 2004.

48. The two-piece assembly was necessary to produce a space between the outer and inner surfaces of the bottle into which the springlike pushtabs could be depressed to unlock the cap. See Thornock et al., "Package Exhibiting Improved Child Resistance."

49. William H. Shecut and Horace H. Day, "Improvement in Adhesive Plasters," U.S. Patent No. 3,965.

50. Johnson & Johnson, "Our Company: Early Years—Our History," http://www.jnj.com/our_company/history/history_section_1.htm. Cf. Charles Panati, *Panati's Extraordinary Origins of Everyday Things* (New York: Harper & Row, 1987), 250–251.

51. "The Band-Aid Brand Story," http://www.bandaid.com/brand_story.shtml.

52. "The Band-Aid Brand Timeline," http://www.bandaid.com/brand_timeline.shtml.

53. Quoted in Caplan, *By Design*, 38.

54. Gregory J. E. Rawlins, *Slaves of the Machine: The Quickening of Computer Technology* (Cambridge: MIT Press, 1997), 95.

55. See, e.g., "Tobacco Heaven," http://www.shopcanada.com/cigars/p1.htm.

56. Michael Brown, "Generation 2023: Talking with the Future," in *Fueling the Future: How the Battle over Energy Is Changing Everything*, ed. Andrew Heintzman and Evan Solomon (Toronto: House of Anansi, 2003), 336.

57. Ronald Ahrens, "RIP Oldsmobile," *Wall Street Journal*, May 4, 2004, editorial page.

58. "New Coke," http://en.wikipedia.org/wiki/New_Coke. Cf. "Knew Coke," http://www.snopes.com/cokelore/newcoke.asp.

59. http://www.uncover.com/coke2.htm.

60. Coca-Cola President and CEO Donald Keough, quoted in "New Coke."

61. "New Coke." Cf. "Knew Coke."

CHAPTER 3: INTANGIBLE THINGS

1. *The Temple. The Church Porch*, Stanza 41. Quoted from Bartlett's, 13th ed.

2. See Henry Petroski, *The Book on the Bookshelf* (New York: Knopf, 1999).

3. John Markoff, "New IMac Makes Debut a Bit Later than Apple Hoped," *New York Times*, September 1, 2004, C9.

4. See http://www.brazzil.com/2004/html/articles/jun04/p111jun04.htm.

5. Donald L. Horowitz, private communications, March 18 and 22, 2004.

6. Ernest Beck and Julie Lasky, "In Iraq, Flag Design, Too, Comes under Fire," *New York Times*, April 29, 2004, F9.

7. Ibid.

8. Ibid.

9. Nadine Post, "Architect's Passion Is Space between Towers, Not Their Height," *Engineering News-Record*, January 15, 1996, 39.

10. See, e.g., "Dr. James Naismith, Inventor of Basketball," http://www.ukans.edu/heritage/graphics/people/naismith.html.

11. See *Sports Illustrated*, April 11, 2001, 26, http://www.duke.edu/~wmp/stuff/duke.html.

12. See http://www.fansonly.com/auto_pdf/p_hotos/s_chools/duke/sports/m-baskbl/auto_pdf/dukerecyearbyyear.

13. See http://www.bbhighway.com/Talk/Coach%20Library/Reviews/Books/Bjarkman_review.asp.

14. Fred Barnes, "One Game at a Time," *Wall Street Journal*, June 9, 2004, D4. See also the book that Barnes was reviewing: Michael Mandelbaum, *The Meaning of Sports: Why Americans Watch Baseball, Football, and Basketball and What They See When They Do* (New York: PublicAffairs, 2004).

15. See http://wyomingathletics.collegesports.com/genrel/wyo-hof.html.

16. "History of Basketball," http://www.planetpapers.com/Assets/3753.php.

17. "Dr. James Naismith."

18. Lisa Jardine, *The Curious Life of Robert Hooke: The Man Who Measured London* (New York: HarperCollins, 2004), 91. Parts of this discussion of Hooke are taken from the author's review of Jardine: "Urban Legend: A Study of the Life and Times of England's Leonardo," *Washington Post Book World*, March 21, 2004, BW04–05.

19. Jardine, *Curious Life*, 97.

20. Quoted in ibid., 231, 232.

21. Quoted in ibid., 227.

22. See Sherwin B. Nuland, *The Doctors' Plague: Germs, Childbed Fever, and the Strange Story of Ignác Semmelweis* (New York: Norton, 2003).

23. Abigail Zuger, "Anatomy Lessons, a Vanishing Rite for Young Doctors," *New York Times*, March 23, 2004, F1, F6.

24. Timothy Rowe, "Anatomy the Old-Fashioned Way," letter to the editor, *New York Times*, May 30, 2004, F4.

25. Nuland, *Doctors' Plague*, 63.

26. Lucian L. Leape, "Error in Medicine," *Journal of the American Medical Association* 232 (December 21, 1994), 1851; Institute of Medicine, Committee on Health Care in America, *To Err Is Human: Building a Safer Health System* (Washington, D.C.: National Academies Press, 1999). Available at http://www.iom.edu/report.asp?id=5575.

27. Quoted in William Jacott, "Medical Errors and Patient Safety: Despite Widespread Attention to the Issue, Mistakes Continue to Occur," *Postgraduate Medicine* (Minneapolis) 114, no. 3 (September 2003), 15.

28. George H. Thomson, "American Bridge Failures: Mechanical Pathology, Considered in Its Relation to Bridge Design," *Engineering*, September 14, 1888, 252–253; also September 21, 294.

29. Carper, "Construction Pathology," 57.

30. Leape, "Error in Medicine," 1851.

31. James Amrhein, quoted in Carper, "Construction Pathology," 57.

32. Lydia Polgreen, "The Pen Is Mightier than the Lock," *New York Times*, September 17, 2004, B1, B6.

33. Lydia Polgreen, "Bicycle Lock Manufacturer to Provide Free Locks to Owners of Easily Unlockable Model," *New York Times*, September 23, 2004, B3.

34. David W. Chen, "Boy, 6, Dies of Skull Injury during M.R.I.," *New York Times*, July 31, 2001, B1.

35. See Nancy G. Leveson and Clark S. Turner, "An Investigation of the Therac-25 Accidents," *Computer*, July 1993, 18–41. See also Steven Casey, *Set Phasers on Stun: And Other True Tales of Design, Technology, and Human Error* (Santa Barbara, Calif.: Aegean Publishing, 1993).

36. For a more detailed description of the case of the Therac-25, see Leveson and Turner, "Therac-25 Accidents."

CHAPTER 4: THINGS SMALL AND LARGE

1. Quoted in Ben Schott, *Schott's Original Miscellany* (New York: Bloomsbury, 2002), 9.

2. Cf. James Deetz, *In Small Things Forgotten: An Archaeology of Early American Life*, rev. and expanded ed. (New York: Anchor Books, 1996).

3. See, e.g., James Nasmyth, *James Nasmyth, Engineer: An Autobiography*, ed. Samuel Smiles (London: John Murray, 1885).

4. Carolyn C. Cooper, "Myth, Rumor, and History: The Yankee Whittling Boy as Hero and Villain," *Technology and Culture* 44 (2003), 85–86.

5. According to *Webster's New International Dictionary*, 1st ed., a "seventy-four" was "an old-time ship of war rated as carrying seventy-four guns."

6. Quoted in Cooper, "Myth, Rumor, and History," 85, from Rev. J. Pierpont, "Whittling—A Yankee Portrait," *United States Magazine* 4 (March 1857), 217.

7. Georgius Agricola, *De Re Metallica*, trans. Herbert Clark Hoover and Lou Henry Hoover (New York: Dover Publications, 1950).

8. See, e.g., Billington, *The Innovators*, 23–29.

9. Cooper, "Myth, Rumor, and History," 93.

10. Derrick Beckett, *Stephensons' Britain* (Newton Abbott, Devon.: David & Charles, 1984), 138.

11. H. J. Hopkins, *A Span of Bridges: An Illustrated History* (Newton Abbott, Devon.: David & Charles, 1970), 159; Sheila Mackay, *The Forth Bridge: A Pictorial History* (Edinburgh: HMSO, 1990), 69.

12. Hopkins, *Span of Bridges*, 159.

13. L.T.C. Rolt, *Isambard Kingdom Brunel* (Hammondsworth, Middlesex: Penguin Books, 1970), 249.

14. Quoted in *New York Times*, August 22, 1923, 15.

15. See, e.g., Mackay, *Forth Bridge*, 112.

16. Dixie W. Golden, "A Man and His Bridge," in *A Golden Gate Jubilee, 1937–1987* (Cincinnati: College of Engineering, University of Cincinnati), 1987.

17. See, e.g., Billington, *The Innovators*, chap. 7; Kenneth Silverman, *Lightning Man: The Accursed Life of Samuel F. B. Morse* (New York: Knopf, 2003), chap. 7.

18. See, e.g., Tom Standage, *The Victorian Internet: The Remarkable Story of the Telegraph and the Nineteenth Century's On-Line Pioneers* (New York: Walker, 1998).

19. Thomas A. Watson, Alexander Graham Bell's assistant, expressed disappointment with Bell's pedestrian first telephone message. Had Bell known he was making history in calling out, "Mr. Watson—Come here—I want to see you," he "would have been prepared with a more resounding and interesting sentence." Watson considered Samuel Morse's "What hath God wrought?" to be "an

example of a 'noble' first message." See Ira Flatow, *They All Laughed . . . From Light Bulbs to Lasers: The Fascinating Stories Behind the Great Inventions that Have Changed Our Lives* (New York: Harper Perennial, 1993), 82n.

20. See, e.g., Whitcomb L. Judson, "Shoe-fastening," U.S. Patent No. 504,037, and Whitcomb L. Judson, "Clasp Locker and Unlocker for Shoes," U.S. Patent No. 504,038. See also Henry Petroski, "On Dating Inventions," *American Scientist*, July–August 1993, 314–318.

21. W. C. Carter, "Umbrella-Stand," U.S. Patent No. 323,397.

22. Simeon S. Post, "Improvement in Iron Bridges," U.S. Patent No. 38,910.

23. "New Miscellaneous Inventions," *Scientific American* 37, no. 9 (September 1, 1877), 138.

24. On mavens, see Malcolm Gladwell, *The Tipping Point: How Little Things Can Make a Big Difference* (New York: Little, Brown, 2002), 59–68.

25. See, e.g., Neil Schlager, ed., *When Technology Fails: Significant Technological Disasters, Accidents, and Failures of the Twentieth Century* (Detroit: Gale Research), 510–517.

26. See, e.g., W. Wayt Gibbs, "Software's Chronic Crisis," *Scientific American*, September 1994, 86.

27. See Bill Wolmuth and John Surtees, "Crowd-related Failure of Bridges," *Proceedings of ICE*, Civil Engineering 156 (2003), 118.

28. "The Great Exhibition Building: Testing the Galleries," *Illustrated London News*, March 1, 1851, 175–176; also February 22, 162.

29. Millennium Bridge Trust, *Blade of Light: The Story of London's Millennium Bridge* (London: Penguin, 2001), 86.

30. Mike Winney, "Dampers to Cut Normandie Vibrations," *New Civil Engineer*, December 15/22, 1994, 5; also January 12, 1995, 5.

31. See, e.g., *Popper Selections*, ed. David Miller (Princeton: Princeton University Press, 1985), 136.

32. James Glanz and Eric Lipton, *City in the Sky: The Rise and Fall of the World Trade Center* (New York: Times Books, 2003), 131.

33. Galileo, *Dialogues Concerning Two New Sciences*, trans. Henry Crew and Alfonso de Salvio (New York: Dover Publications, [1954]), 2–6.

CHAPTER 5: BUILDING ON SUCCESS

1. *Elective Affinities*, trans. James Anthony Froude, Book I, ch. 9. Quoted from *Bartlett's*, 16th ed., 349.

2. "The Failure of Buildings," *The American Architect and Building News* 32, no. 798 (April 11, 1891), 28–29.

3. Louis Torres, *"To the immortal name and memory of George Washington": The United States Army Corps of Engineers and the Construction of the Washington Monument* (Washington, D.C.: Office of the Chief of Engineers, n.d.), 106–107.

4. Quoted in ibid., 107.

5. Edwin G. Burrows and Mike Wallace, *Gotham: A History of New York City to 1898* (New York: Oxford University Press, 1999), 63–64.

6. Neal Bascomb, *Higher: A Historic Race to the Sky and the Making of a City* (New York: Doubleday, 2003), 9.

7. Sarah Bradford Landau and Carl W. Condit, *Rise of the New York Skyscraper, 1865–1913* (New Haven, Conn.: Yale University Press, 1996), 381–382, 389–90, 384, 386.

8. Food shops have been especially prone to whimsical design. For a picture of the Clam Box, "a gray wooden shack shaped like the cardboard container in which fried clams are traditionally served," see R. W. Apple, Jr., "Even the Body Politic Has to Eat," *New York Times*, July 21, 2004, F1, F6. See also, e.g., Robert Venturi, Steven Izenour, and Denise Scott Brown, *Learning from Las Vegas—Revised Edition: The Forgotten Symbolism of Architectural Form* (Cambridge: MIT Press, 1977).

9. "No Picnic for Designers," *Modern Steel Construction*, May 1998, 44–48. See also "Robert McG. Thomas Jr., "David Longaberger, Basket Maker, Dies at 64," *New York Times*, March 22, 1999, obituary page.

10. Quoted in Bascomb, *Higher*, 150.

11. Gustave Eiffel, quoted in "The Tower Stirs Debate & Controversy," http://www.tour-eiffel.fr/teiffel/uk/documentation/dossiers/page/debats.html; translated from *Le Temps* (Paris), February 14, 1887.

12. Bascomb, *Higher*, 102–103.

13. William Foreman, "Taipei 101 Skyscraper Deemed Tallest," Associated Press Online, October 8, 2004.

14. Herbert Muschamp, "Skyscraping around the Urban World," *New York Times*, July 16, 2004, E25, E29.

15. "A Super Engineer's Skyscraper," *Structural Engineer*, May 2004, 42.

16. See, e.g., Fazlur Khan, "The John Hancock Center," *Civil Engineering*, October 1967, 38–42. See also Yasmin Sabina Khan, *Engineering Architecture: The Vision of Fazlur R. Khan* (New York: Norton, 2004), 69–70.

17. "A Super Engineer's Skyscraper."

18. Glanz and Lipton, *City in the Sky*, 139–140.

19. See, e.g., Matthys Levy and Mario Salvadori, *Why Buildings Fall Down: How Structures Fail* (New York: Norton, 1992), 197–205.

20. Ibid., 202.

21. Joe Morgenstern, "The Fifty-Nine Story Crisis," *New Yorker*, May 29, 1995, 45–53.

22. See, e.g., Dennis C. K. Poon et al., "Reaching for the Sky," *Civil Engineering*, 54–61, 72.

23. Foreman, "Taipei 101 Skyscraper."

24. Muschamp, "Skyscraping around the Urban World," *New York Times*, July 16, 2004, E25, E29.

25. David Littlejohn, "It's a Pickle, It's a Pineapple—It's a Brilliant New Skyscraper," *Wall Street Journal*, July 13, 2004, D8.

26. Ibid.

27. James Barron, "Queen Mary 2 Is Back, This Time Bringing Better Bathrooms," *New York Times*, July 6, 2004, B8.

CHAPTER 6: STEPPING-STONES TO SUPER-SPANS

1. "Metaphors of a Magnifico," *The Collected Poems of Wallace Stevens* (New York: Knopf, 1954), 19.

2. Anne Raver, "In the Lair of a Tender Giant," *New York Times*, September 25, 2003, D9.

3. See, e.g., Mark Lehner, *The Complete Pyramids* (London: Thames and Hudson, 1997).

4. See, e.g., Steven M. Richman, *The Bridges of New Jersey: Portraits of Garden State Crossings* (New Brunswick, N.J.: Rutgers University Press, 2005), 2.

5. Aristotle, "Mechanical Problems," question 17.

6. Hopkins, *Span of Bridges*, 18.

7. David de Haan, "The Iron Bridge—How Was It Built?" http://www.bbc.co.uk/history/society_culture/industrialisation/iron_bridge_01.shtml (October 13, 2005).

8. See, e.g., Derrick Beckett, *Stephensons' Britain* (Newton Abbot, Devon.: David & Charles, 1984), 124, fig. 29.

9. Sir Alfred Pugsley, *The Theory of Suspension Bridges* (London: Edward Arnold, 1957), 1–2.

10. Hopkins, *Span of Bridges*, 83, 85.

11. See, e.g., Billington, *The Innovators*, 149.

12. James Sutherland, "Iron Railway Bridges," in *Robert Stephenson—The Eminent Engineer*, ed. Michael R. Bailey (Aldershot, Hants.: Ashgate, 2003), 318. Cf. Cyril Stapley Chettoe, Norman Davey, and George Robinson Mitchell, "The Strength of Cast-Iron Girder Bridges," *Journal of the Institution of Civil Engineers* 22 (1944), 246.

13. See, e.g., Peter R. Lewis and Colin Gagg, "Aesthetics Versus Function: The Fall of the Dee Bridge, 1847," *Interdisciplinary Science Reviews* 29 (2004), 177–191.

14. See, e.g., T. Martin and I. A. MacLeod, "The Tay Bridge Disaster—A Reappraisal Based on Modern Analysis Methods," *Proceedings of the Institution of Civil Engineers*, Civil Engineering 108 (1995), 77–83.

15. Peter R. Lewis and Ken Reynolds, "Forensic Engineering: A Reappraisal of the Tay Bridge Disaster," *Interdisciplinary Science Reviews* 27 (2002): 287–298. See also Martin and MacLeod, "Tay Bridge Disaster."

16. Martin and MacLeod, "Tay Bridge Disaster," 83.

17. Quoted in Jim Crumley, *The Forth Bridge* (Grantown-on-Spey, Moray, Scotland: Colin Baxter Photography, 1999), 9.

18. Quoted in Sutherland, "Iron Railway Bridges," 331.

19. For a contemporary description of the Forth Bridge project, see B. Baker, "Bridging the Firth of Forth," *Engineering*, July 29, 1887, 116; August 5, 148; August 12, 170–171; August 19, 210; August 26, 238.

20. For a recent review of the Quebec Bridge failure, see, e.g., William D. Middleton, *The Bridge at Quebec* (Bloomington: Indiana University Press, 2001).

21. David B. Steinman and Sara Ruth Watson, *Bridges and Their Builders*, rev. and expanded ed. (New York: Dover Publications, 1957), 243.

22. The film has been available from the Department of Civil Engineering, University of Washington.

23. For post-Tacoma Narrows suspension bridges, see Richard Scott, *In the Wake of Tacoma: Suspension Bridges and the Quest for Aerodynamic Stability* (Reston, Va.: ASCE Press, 2001).

24. Peter R. Lewis, e-mail messages to author, August 8–9, 2004.

CHAPTER 7: THE HISTORICAL FUTURE

1. *Endymion*, preface. Quoted from *Bartlett's*, 16th ed., 414.

2. Letter to James Hessey, October 8, 1818. Quoted from *Bartlett's*, 16th ed., 418.

3. Diane Vaughan, *The Challenger Launch Decision: Risky Technology, Culture, and Deviance at NASA* (Chicago: University of Chicago Press, 1996), chap. 3.

4. Richard Feynman, quoted in ibid., 274.

5. Quoted in Warren E. Leary, "Debating the Real Price of Space Bargains," *New York Times*, May 9, 2000, F3.

6. James Oberg, "Why the Mars Probe Went Off Course," *IEEE Spectrum*, December 1999, 34.

7. Liam P. Sarsfield, quoted in ibid.

8. Robert Lee Hotz, "Butterfly on a Bullet: Firing Point-Blank at NASA's Illusions," *Los Angeles Times*, December 25, 2003, A-1. For background, see Vaughan, *Challenger Launch Decision*.

9. See, among countless newspaper reports, Warren E. Leary, "The Old Shuttle, New Again," *New York Times*, July 12, 2005, F1, F4–F5; John Schwartz, "NASA Is Said to Loosen Risk Standards for Shuttle," April 22, A1, A21; John Schwartz and Warren E. Leary, "Shuttle Repairs to Be Tried in Spacewalk," August 2, A1, A19; John Schwartz and Warren E. Leary, "Shuttle Glides to Safe Landing; Problems Ahead," August 10, A1, A16.

10. See John A. Roebling, "Remarks on Suspension Bridges, and on the Comparative Merits of Cable and Chain Bridges," *American Railroad Journal, and Mechanics' Magazine* n.s., 6 (1841), 193–196.

11. Some of this material, in a slightly different form, appeared first in Henry Petroski, "Look First to Failure," *Harvard Business Review*, October 2004, 18–20.

12. P. G. Sibly and A. C. Walker, "Structural Accidents and Their Causes," *Proceedings of the Institution of Civil Engineers*, part 1, 62 (1977), 191–208.

13. Henry Petroski, "Predicting Disaster," *American Scientist*, March–April 1993, 110–113.

14. Millennium Bridge Trust, *Blade of Light*.

15. Mary Blume, "Pont Solferino: Water under a Troubled Bridge," *International Herald Tribune*, at http://www.iht.com/IHT/MB/00/mb072900.html.

16. See, e.g., "World's Longest Cable-Stayed Bridge to be Built in China," *Structural Engineer*, May 2004, 11.

17. Terry Stephens, "Special Dampers May Shake Up Engineering Field," November 20, 2003, at http://www.djc.com/news/ae/11151055.html.

18. Quoted from Schlager, ed., *When Technology Fails*, 210.

19. See, e.g., "Palau Files Suit in Bridge Collapse," *Civil Engineering*, July 1997, 12–13.

20. Bridget McCrea, "Florida Crossing Work Halts as Contractor Faces Heat," *Engineering News-Record*, June 21, 2004, 19–20.

21. Although a new cohort is "born" (graduates from college) every year, it does not follow that a fresh crop of engineers joins a particular technological "family" at that same rate. When a sector of an industry grows rapidly, it tends to hire new (young) engineers at an equally rapid pace—until business levels off. When hard times come, an entire industry may cease hiring for a prolonged period, making do with an aging population of engineers and not replacing retiring ones with young ones. Only when business picks up does hiring resume. As a result, there can develop a "generational gap" between the experienced and inexperienced engineers. One observer noticed in 1981 that "everyone on the R&D side" of the oil industry "was either under 30 or over 50 years old" (Alex Pavlak, e-mail to author, August 7, 2005).

22. Sibly and Walker, "Structural Accidents," 208. When *Mechanical Engineering* magazine asked its readers how their company passed on "knowledge of its senior engineers to younger colleagues," responses revealed that about a quarter had "no inherent cultural expectation" that it be done at all. About four out of ten engineers reported doing so informally, "at the proverbial water cooler." Only 13

percent of the respondents indicated that their company made time to do so "now and then at staff meetings," and only 22 percent said they had "formal guidelines and/or tutorials." See *Mechanical Engineering*, January 2005, 10; see also www.memagazine.org.

23. Robert Lee Hotz, "Butterfly on a Bullet: The Fate of a Wing Shaped by Politics," *Los Angeles Times*, December 24, 2003, A-1. All installments of this six-part series appeared under the same main title, but with different subtitles, on December 21, 22, 23, 24, 25, and 26, 2003, all beginning on p. A-1.

24. Paul Dimotakis, quoted in ibid., December 24.

25. Ed Friedman, quoted in Leary, "Debating the Real Price."

26. Calvin Sims, "Angst at Japan Inc.," *New York Times*, December 3, 1999, C1, C6.

27. Some of this material appeared first, in slightly different form, in Henry Petroski, "Past and Future Failures," *American Scientist*, November–December 2004, 500–504.

28. Franco Moretti, "Graphs, Maps, Trees: Abstract Models for Literary History—1," *New Left Review* 24 (2003), 67–93. See also Emily Eakin, "Studying Literature by the Numbers," *New York Times*, January 10, 2004, B9.

29. Thomas S. Kuhn, *The Structure of Scientific Revolutions* (Chicago: University of Chicago Press, 1962).

30. Moretti, "Graphs, Maps, Trees."

31. Quoted in Sutherland, "Iron Railway Bridges," 313.

32. Quoted in ibid., 315.

33. On the Britannia Bridge as a failure, see Henry Petroski, *Design Paradigms: Case Histories of Error and Judgment in Engineering* (Cambridge: Cambridge University Press, 1994), chap. 7.

34. Quoted in R. R. Whyte, ed., *Engineering Progress through Trouble* (London: Institution of Mechanical Engineers, 1975), v.

35. In *On Medical Education*. Quoted from *Bartlett's*, 13th ed.

36. See, e.g., Petroski, *Design Paradigms*.

37. J. A. L. Waddell, *Bridge Engineering*, 2 vols. (New York: Wiley, 1916).

38. "Why Not Have All Designs Checked by Outside Experts?" *Engineering News-Record*, November 22, 1917, 979–980.

39. J. A. L. Waddell, quoted in ibid.

40. Paul Sibly, "The Prediction of Structural Failures," Ph.D. Thesis, University of London, 1977.

41. Quoted in "Why Not Have All Designs Checked," 980.

42. See, e.g., Henry Petroski, *Engineers of Dreams: Great Bridge Builders and the Spanning of America* (New York: Knopf, 1995), 296–300.

43. P. S. Bulson, J. B. Caldwell, and R. T. Severn, eds., *Engineering Structures: Developments in the Twentieth Century: A Collection of Essays to Mark the Eightieth Birthday of Sir Alfred Pugsley* (Bristol: University of Bristol Press, 1983).

44. Roebling, "Remarks on Suspension Bridges."

45. Pugsley, *Theory of Suspension Bridges*, 6.

46. See, e.g., Leon S. Moisseiff and Frederick Lienhard, "Suspension Bridges under the Action of Lateral Forces," *Transactions of the American Society of Civil Engineers* 98 (1933), 1080–1095.

47. Ibid., 1–10. See also David B. Steinman, "Suspension Bridges: The Aerodynamic Problem and Its Solution," *American Scientist*, July 1954, 397–438, 460.

48. H. Kit Miyamoto and Robert D. Hanson, "Seismic Dampers: State of the Applications," *Structure*, July 2004, 16–18.

49. Robert Byrne, *Skyscraper* (New York: Atheneum, 1984).

50. Joe Morgenstern, "The Fifty-Nine Story Crisis," *New Yorker*, May 29, 1995, 45–53.

51. See, e.g., Gretchen Ruethling, "36 Are Hurt as Fire Damages Bank Headquarters in Chicago," *New York Times*, December 8, 2004, national edition, A18.

52. See, e.g., Schlager, ed., *When Technology Fails*, 307–312.

53. Peter Reina, "Focus on Construction of Columns at Airport,"

Engineering News-Record, May 31, 2004, 10–11. See also sidebar, "Moving Far Beyond Airports," 11.

54. Craig S. Smith, "New Cracks Stop Search at Terminal after Collapse," *New York Times*, May 25, 2004, 10.

55. See Aileen Cho, "Planned, Collapsed Terminals Featured at Peer Review," *Engineering News-Record*, September 27, 2004, 16.

56. Peter Reina, "Airport Roof Failure Blamed on Process," *Engineering News-Record*, February 21, 2005, 10–11; "More Team Building," editorial, ibid., 56.

57. Craig S. Smith, "Architect Starts Study of Failure in Paris Airport," *New York Times*, May 26, 2004, A13.

58. Reina, "Focus on Construction," 11.

59. Joseph Kahn, "A Glass Bubble That's Bringing Beijing to a Boil," *New York Times*, June 15, 2004, A1, A12. See also Janice Tuchman, Peter Reina, and Andrea Ding Kemp, "Beijing's National Grand Theater Transforms the Cityscape," *Engineering News-Record*, November 29, 2004, 22–27.

60. Moisseiff's obituary in the *Transactions of the American Society of Civil Engineers* did not mention the Tacoma Narrows Bridge. See vol. 111 (1946), 1509–1512.

61. "Paris Airport Collapse Should Not Stifle Innovation," *Engineering News-Record*, May 31, 2004, 56.

62. Sir Alfred Pugsley, *The Safety of Structures* (London: Edward Arnold, 1966), 2.

63. Quoted in Alfred Greenville Pugsley, "Concepts of Safety in Structural Engineering," *Journal of the Institution of Civil Engineers* 36 (1951), 29.

64. "Charley's Rampage Also Shakes Construction's Trust," *Engineering News-Record*, August 23, 2004, 96. See also "New Florida Codes Bring Mixed Success," ibid., 8–9.

65. Eric Lipton, "After 100 Years, a New Rule Book for New York," *New York Times*, May 17, 2004, A21.

66. See, e.g., *Engineering News-Record*, June 7, 2004.

67. Nadine M. Post, "Skyscrapers' Supporters Infuriated by Fire Fearmongers," *Engineering News-Record,* June 7, 2004, 48–54.

68. Milt Freudenheim, "Many Hospitals Resist Computerized Patient Care," *New York Times*, April 6, 2004, C1, C6. See also Institute of Medicine, Committee on Health Care in America, *To Err Is Human.*

69. W. E. Deming, quoted in Lucian L. Leape, "Error in Medicine," *Journal of the American Medical Association* 272 (December 21, 1994), 1851.

70. Ibid., 1852.

INDEX

Italicized page numbers refer to illustrations and their captions.